SCALAR AND ASYMPTOTIC SCALAR DERIVATIVES

Theory and Applications

Springer Optimization and Its Applications

VOLUME 13

Aims and Scope
Optimization has been expanding in all directions at an astonishing rate during the last few decades. New algorithmic and theoretical techniques have been developed, the diffusion into other disciplines has proceeded at a rapid pace, and our knowledge of all aspects of the field has grown even more profound. At the same time, one of the most striking trends in optimization is the constantly increasing emphasis on the interdisciplinary nature of the field. Optimization has been a basic tool in all areas of applied mathematics, engineering, medicine, economics and other sciences.

The series *Optimization and Its Applications* publishes undergraduate and graduate textbooks, monographs and state-of-the-art expository works that focus on algorithms for solving optimization problems and also study applications involving such problems. Some of the topics covered include nonlinear optimization (convex and nonconvex), network flow problems, stochastic optimization, optimal control, discrete optimization, multi-objective programming, description of software packages, approximation techniques and heuristic approaches.

SCALAR AND ASYMPTOTIC SCALAR DERIVATIVES

Theory and Applications

By

GEORGE ISAC
Royal Military College of Canada, Kingston, Ontario, Canada

SÁNDOR ZOLTÁN NÉMETH
University of Birmingham, Birmingham, United Kingdom

 Springer

Authors
George Isac
Royal Military College of Canada
Department of Mathematics
Kingston ON K7K 7B4
STN Forces
Canada
gisac@juno.com

Sándor Zoltán Németh
The University of Birmingham
School of Mathematics
The Watson Building
Edgbaston
Birmingham
B15 2TT

ISBN: 978-1-4419-4484-9   e-ISBN: 978-0-387-73988-5

Printed on acid-free paper

9 8 7 6 5 4 3 2 1

springer.com

The sage gives without reservation.
He offers all to others, and his life
is more abundant. He helps all men alike,
and his life is more exuberant.

(Lao Zi: *Truth and Nature*)

Preface

This book is devoted to the study of scalar and asymptotic scalar derivatives and their applications to the study of some problems considered in nonlinear analysis, in geometry, and in applied mathematics.

The notion of a scalar derivative is due to S. Z. Németh, and the notion of an asymptotic scalar derivative is due to G. Isac. Both notions are recent, never considered in a book, and have interesting applications. About applications, we cite applications to the study of complementarity problems, to the study of fixed points of nonlinear mappings, to spectral nonlinear analysis, and to the study of some interesting problems considered in differential geometry and other applications.

A new characterization of monotonicity of nonlinear mappings is another remarkable application of scalar derivatives.

A relation between scalar derivatives and asymptotic scalar derivatives, realized by an inversion operator is also presented in this book. This relation has important consequences in the theory of scalar derivatives, and in some applications. For example, this relation permitted us a new development of the method of exceptional family of elements, introduced and used by G. Isac in complementarity theory.

Now, we present a brief description of the contents of this book.

Chapter 1 is dedicated to the study of scalar derivatives in Euclidean spaces. In this chapter we explain the reason for introducing scalar derivatives as good mathematical tools for characterizing important properties of functions from $\mathbb{R}^n$ to $\mathbb{R}^n$. In order to avoid some difficulties, we consider only upper and lower scalar derivatives which are extensions to vector functions of Dini derivatives. We consider also the case when lower and upper scalar derivatives coincide. This is a strong restriction and we show that for $n = 2$ the existence of a single-valued scalar derivative is strongly related to complex differentiability. The lower and upper scalar derivatives are also used to characterize convexity like notions.

Chapter 2 essentially has two parts. In the first part we present the notion of the asymptotic derivative and some results related to this notion and in the second part we introduce the notion of the asymptotic scalar derivative. The results presented in the first part are necessary for understanding the notions given in the second part. It is known that the notion of the asymptotic derivative was introduced by the Russian school, in particular by M. A. Krasnoselskii, under the name of asymptotic linearity. The main goal of this chapter is to present the notion of the asymptotic scalar derivative and some of its applications.

Chapter 3 presents the scalar derivatives in Hilbert spaces and several results and properties are given. We note that in this chapter we give the definitions of scalar derivatives of rank p, named briefly for $p = 2$, scalar derivatives. We also put in evidence the fact that the case $p = 1$ is strongly related to the notion of submonotone mapping, introduced in 1981 by J. E. Spingarn and studied in 1997 by P. Georgiev. Several new results related to computation of the scalar derivative and some interesting relations with skew-adjoint operators are also presented. The scalar derivatives are used to characterize the monotonicity of mappings in Hilbert spaces. Many of the formulae presented in this chapter arise from applications such as *fixed point* theorems, surjectivity theorems, integral equations, and complementarity problems, among others.

Chapter 4 contains the extension of the theory of scalar derivatives to Banach spaces. This extension is based on the notion of the semi-inner product in Lumer's sense. The notion of scalar derivatives defined in this case is applied to fixed point theory, to the study of solvability of integral equations, of variational inequalities, and of complementarity problems.

Chapter 5 is dedicated to a generalization of the notion of Kachurovskii–Minty–Browder monotonicity to Riemannian manifolds and to realize this we introduce the notion of the geodesic monotone vector field. The geodesic convexity for mappings is also considered. For a global example of monotone vector fields we consider Hadamard manifolds (complete, simply connected Riemannian manifolds with nonpositive sectional curvature). Analyzing the existence of geodesic monotone vector fields, we prove that there are no strictly geodesic monotone vector fields on a Riemannian manifold that contain a closed geodesic. We note that many results presented in this chapter are based on a generalization to Riemannian manifolds of scalar derivatives studied in the previous chapters. The nongradient type monotonicity on Riemannian manifolds is considered for the first time in a book.

This book is the first book dedicated to the study of scalar and asymptotic scalar derivatives and certainly new developments related to these notions are possible.

It is impossible to finish this preface without giving many thanks to the people who spent their time developing the open source tools (operating system, window manager, and software) that were essential for writing this book, greatly reducing the time and energy spent in word processing. These open source tools are: the Linux and FreeBSD operating systems, the Ratpoison window manager, the LaTeX word processing language, and the VIM and Bluefish editors.

We are grateful to the reviewers for their valuable comments and suggestions. Taking them into consideration has greatly improved the quality and presentation of the book.

To conclude, we would like to say that we very much appreciated the excellent assistance offered to us by the staff of Springer Publishers.

Canada   George Isac
Birmingham, UK   Sándor Zoltán Németh

Contents

Chapter 1

Scalar Derivatives in Euclidean Spaces

1.1 Scalar Derivatives of Mappings in Euclidean Spaces

The behaviour of the scalar product $\langle f(x) - f(y), x - y \rangle$ (with $f : \mathbb{R}^n \to \mathbb{R}^n$ and $\langle\,.\,,\,.\,\rangle$ the usual scalar product in $\mathbb{R}^n$) when x and y run over $\mathbb{R}^n$ is a good tool in characterizing important properties of f. If f is bounded, then this product converges to 0 for $x \to y$. Therefore it cannot be used in obtaining a local characterization. Hence it is natural to consider at y limits of the expressions of the form $\langle f(x) - f(y), x - y \rangle / \langle x - y, x - y \rangle$ for $x \to y$. Thus we arrive naturally at a notion that we call the scalar derivative. It is in general a multivalued mapping from $\mathbb{R}^n$ to $\mathbb{R}$ even if f is linear.

In order to avoid the difficulties in considering multifunctions we only consider so-called upper and lower scalar derivatives, which are extensions to vector functions of the Dini derivatives.

We consider mostly the case when lower and upper scalar derivatives coincide. This restriction is a very strong one. In Section 1.1.3 it is shown that for $n = 2$ the existence of a single-valued scalar derivative is strongly related to the complex differentiability. In Section 1.1.4 we consider various examples and counterexamples. Lower and upper scalar derivatives can be used in characterizing the monotone operators in the way this is done in Section 1.1.5.

Convex functionals have as gradients monotone operators. Hence the scalar derivative can also be used to characterize convexity like notions. Thus Propositions 2.1 and 2.2 in Karamardian and Schaible [1990] together with the results in our Section 1.1.5 give some characterizations of convex and strictly convex functionals.

We have defined the notion of scalar derivative having in mind Minty's monotonicity notion [Minty, 1962]. To simplify the notations, in this chapter a monotone mapping (strictly monotone mapping) f will be called increasing (strictly increasing). If $-f$ is monotone (strictly monotone), then f will be called decreasing (strictly decreasing).

1.1.1   Some Basic Results Concerning Skew-Adjoint Operators

DEFINITION 1.1 *Consider the operator $f : \mathbb{R}^n \to \mathbb{R}^n$. It is called* increasing *(decreasing) if for any x and y in $\mathbb{R}^n$ one has*

$$\langle f(x) - f(y),\, x - y \rangle \geq 0 \ (\leq 0).$$

If

$$\langle f(x) - f(y),\, x - y \rangle > 0 \ (< 0)$$

whenever $x \neq y$, then f is called strictly increasing *(strictly decreasing).*

DEFINITION 1.2 *The linear operator $A : \mathbb{R}^n \to \mathbb{R}^n$ is called* skew-adjoint *if for any x and y in $\mathbb{R}^n$ the relation $\langle Ax,\, y \rangle + \langle Ay,\, x \rangle = 0$ holds.*

THEOREM 1.3 *If $A : \mathbb{R}^n \to \mathbb{R}^n$ is linear, then the following statements are equivalent.*

1. *A is skew-adjoint.*
2. *$\langle Ax - Ay,\, x - y \rangle = 0$ for any $x,\, y \in \mathbb{R}^n$.*
3. *Taking an arbitrary orthonormal basis in $\mathbb{R}^n$, A can be represented by a matrix $A = (a_{ij})_{i,j=1,\dots,n}$ such that $a_{ij} = -a_{ji} \ \forall i,\, j \in \{1, 2, \dots, n\}$.*

Proof. $1 \Rightarrow 2$ Take x and y arbitrarily in $\mathbb{R}^n$. By the definition of the skew-adjoint operator A we have $\langle Ax,\, y \rangle + \langle Ay,\, x \rangle = 0$. Put $y = x$. Then $\langle Ax,\, x \rangle = 0$ for arbitrary x in $\mathbb{R}^n$. Whence we also have $\langle Ax - Ay,\, x - y \rangle = 0$ by the linearity of A. The implication $2 \Rightarrow 1$ can be shown similarly.

The equivalence $1 \Leftrightarrow 3$ is obvious. $\square$

REMARK 1.1

1. *There exist injective skew-adjoint operators. For instance, the operators represented by the matrices*

$$A = \begin{pmatrix} 0 & 1 \\ -1 & 0 \end{pmatrix} \qquad \text{for } n = 2$$

   and

$$A = \begin{pmatrix} 0 & -1 & 0 & -1 \\ 1 & 0 & 1 & 0 \\ 0 & -1 & 0 & 0 \\ 1 & 0 & 0 & 0 \end{pmatrix} \qquad \text{for } n = 4$$

   are injective.

2. *If n is odd, then there is no injective skew-adjoint operator in $\mathbb{R}^n$. Indeed let A be the matrix corresponding to an skew-adjoint operator. Let the superscript T denote transposition. Then $A^T = -A$ and hence $\det A = -\det A$ this means that $\det A = 0$.*

THEOREM 1.4 *Consider the operator $F : \mathbb{R}^n \to \mathbb{R}^n$. The following assertions are equivalent.*

1. $\langle F(x) - F(y), x - y \rangle = 0, \forall x, y \in \mathbb{R}^n$.

2. *F is an affine operator with a skew-adjoint linear term.*

Proof. Suppose that 1 holds. Put $f(x) = F(x) - F(0)$ for x in $\mathbb{R}^n$. Then $f(0) = 0$ and $\langle f(x) - f(y), x - y \rangle = 0 \; \forall x, y \in \mathbb{R}^n$. Let x be arbitrary in $\mathbb{R}^n$ and $y = 0$. Then $\langle f(x), x \rangle = 0 \; \forall x \in \mathbb{R}^n$. The above relation also yields

$$\langle f(x), x \rangle - \langle f(x), y \rangle - \langle f(y), x \rangle + \langle f(y), y \rangle = 0, \;\; \forall x, y \in \mathbb{R}^n$$

and hence

$$\langle f(x), y \rangle + \langle f(y), x \rangle = 0, \;\; \forall x, y \in \mathbb{R}^n.$$

Put $x = \lambda x_1 + \mu x_2$ with arbitrary x_1 and x_2 in $\mathbb{R}^n$. Then

$$\langle f(\lambda x_1 + \mu x_2), y \rangle = -\langle f(y), \lambda x_1 + \mu x_2 \rangle = -\lambda \langle f(y), x_1 \rangle - \mu \langle f(y), x_2 \rangle$$
$$= \lambda \langle f(x_1), y \rangle + \mu \langle f(x_2), y \rangle,$$

wherefrom

$$\langle f(\lambda x_1 + \mu x_2) - \lambda f(x_1) - \mu f(x_2), y \rangle = 0$$

for any x_1, x_2 and y in $\mathbb{R}^n$ and any λ, μ in $\mathbb{R}$, wherefrom we have the linearity of f. Because $\langle f(x) - f(y), x - y \rangle = 0$, for any x, y in $\mathbb{R}^n$, f is also skew-adjoint. Thus $F(x) = f(x) + F(0)$ and hence it is indeed affine with a skew-adjoint linear term.

The implication $2 \Rightarrow 1$ is obvious. $\square$

1.1.2 The Scalar Derivative and its Fundamental Properties

DEFINITION 1.5 *Consider the operator $f : \mathbb{R}^n \to \mathbb{R}^n$. If the limit*

$$\lim_{x \to x_0} \frac{\langle f(x) - f(x_0), x - x_0 \rangle}{\|x - x_0\|^2} =: f^{\#}(x_0) \in \mathbb{R}$$

exists (here $\|x - x_0\|^2 = \langle x - x_0, x - x_0 \rangle$), then it is called the scalar derivative *of the operator f in x_0. In this case f is said to be* scalarly differentiable *at x_0. If $f^{\#}(x)$ exists for every x in $\mathbb{R}^n$, then f is said to be* scalarly differentiable on *$\mathbb{R}^n$, with the* scalar derivative *$f^{\#}$.*

It follows from this definition that both the set of operators scalarly differentiable in x_0, and the set of operators scalarly differentiable on $\mathbb{R}^n$ form linear spaces.

DEFINITION 1.6 *Consider the operator* $f : \mathbb{R}^n \to \mathbb{R}^n$. *The limit*

$$\underline{f}^{\#}(x_0) := \liminf_{x \to x_0} \frac{\langle f(x) - f(x_0),\, x - x_0 \rangle}{\|x - x_0\|^2}$$

is called the lower scalar derivative *of* f *at* x_0. *Taking* $\limsup$ *in place of* $\liminf$ *we can define the* upper scalar derivative $\overline{f}^{\#}(x_0)$ *of* f *at* x_0 *similarly.*

THEOREM 1.7 *The linear operator* $A : \mathbb{R}^n \to \mathbb{R}^n$ *is scalarly differentiable on* $\mathbb{R}^n$ *if and only if it is of the form* $A = B + cI_n$ *with* B *skew-adjoint linear operator,* I_n *the identity of* $\mathbb{R}^n$, *and* c *a real number.*

Proof. Let us suppose that A is scalarly differentiable in $x_0 \in \mathbb{R}^n$. Then

$$A^{\#}(x_0) = \liminf_{x \to x_0} \frac{\langle Ax - Ax_0,\, x - x_0 \rangle}{\|x - x_0\|^2} = \liminf_{h \to 0} \frac{\langle Ah,\, h \rangle}{\|h\|^2} = A^{\#}(0).$$

Take $h = \lambda x$ with $x \in \mathbb{R}^n$ and $\lambda > 0$. Then

$$A^{\#}(0) = \liminf_{\lambda \downarrow 0} \frac{\langle A\lambda x,\, \lambda x \rangle}{\|\lambda x\|^2} = \frac{\langle Ax,\, x \rangle}{\|x\|^2}.$$

That is, $\langle Ax,\, x \rangle / \|x\|^2 = c = A^{\#}(0)$. Accordingly,

$$\langle (A - cI_n)x,\, x \rangle = 0,\ \forall x \in \mathbb{R}^n.$$

This means that $B = A - cI_n$ is a skew-adjoint linear operator and hence A has the representation given in the theorem. Obviously, every $A = B + cI_n$ with B a skew-adjoint linear operator has the scalar derivative c at every point of $\mathbb{R}^n$. $\square$

THEOREM 1.8 *Suppose that* $f : \mathbb{R}^n \to \mathbb{R}^n$, $f = (f_1, \cdots, f_n)$ *is scalarly differentiable in* x_0. *Then for every* $i \in \{1, \ldots, n\}$ *there exists the partial derivative*

$$\frac{\partial f_i(x_0)}{\partial x^i} \text{ and } \frac{\partial f_1(x_0)}{\partial x^1} = \cdots = \frac{\partial f_n(x_0)}{\partial x^n} = f^{\#}(x_0).$$

Proof. If we consider $x = (x_0^1, \ldots, x^i, \ldots, x_0^n)$ and $x_0 = (x_0^1, \ldots, x_0^n)$, by letting $x \to x_0$, we obtain that

$$\frac{\partial f_i(x_0)}{\partial x^i} = f^{\#}(x_0).$$

$\square$

THEOREM 1.9 *Suppose that $f : \mathbb{R}^n \to \mathbb{R}^n$ is differentiable in x_0 and scalarly differentiable in x_0. Then we have for the differential $df(x_0)$ of f at x_0 the relation*

$$df(x_0) = B + f^{\#}(x_0)I_n,$$

with $B : \mathbb{R}^n \to \mathbb{R}^n$ linear and skew-adjoint .

Proof. Let $t \in \mathbb{R}^n$ be given. Then

$$f^{\#}(x_0) = \frac{1}{\|t\|^2} \liminf_{\lambda \downarrow 0} \left\langle \frac{f(x_0 + \lambda t) - f(x_0)}{\lambda}, \ t \right\rangle = \frac{1}{\|t\|^2} \langle df(x_0)(t), \ t \rangle,$$

wherefrom $\langle (df(x_0) - f^{\#}(x_0)I_n)(t), \ t \rangle = 0, \forall t \in \mathbb{R}^n$, that is,

$$B = df(x_0) - f^{\#}(x_0)I_n$$

is linear and skew-adjoint. $\square$

REMARK 1.2

1. *The theorem holds for the Gateaux differential $\delta f(x_0)$ in place of $df(x_0)$. The differentiability condition is often used next and hence we state the theorem for this stronger condition.*

2. *If we denote by $f'(x_0)$ the Jacobi matrix of f at x_0 in some coordinate representation and the symbols B and I_n stand for matrices of the corresponding operators, then our relation becomes*

$$f'(x_0) = B + f^{\#}(x_0)I_n.$$

THEOREM 1.10 *Suppose that $f : \mathbb{R}^n \to \mathbb{R}^n$, $f = (f_1, \dots, f_n)$ is differentiable in x_0. Then the following statements are equivalent.*

1. *f is scalarly differentiable in x_0;*

2. (a) $\dfrac{\partial f_1(x_0)}{\partial x^1} = \cdots = \dfrac{\partial f_n(x_0)}{\partial x^n}$;

   (b) $\dfrac{\partial f_i(x_0)}{\partial x^j} = -\dfrac{\partial f_j(x_0)}{\partial x^i}, \ \forall i, j \in \{1, \dots, n\}, \ i \neq j.$

Condition 2 is called the *Cauchy–Riemann relation* at x_0.

Proof. $1 \Rightarrow 2$ By Remark 2 one has

$$f'(x_0) = B + f^{\#}(x_0)I_n,$$

where B is a skew-symmetric matrix and $f'(x_0)$ is the Jacobi matrix of f at x_0. Because from the above relation

$$
B = \begin{pmatrix}
\dfrac{\partial f_1(x_0)}{\partial x^1} - f^{\#}(x_0) & \dfrac{\partial f_1(x_0)}{\partial x^2} & \cdots & \dfrac{\partial f_1(x_0)}{\partial x^n} \\[3mm]
\dfrac{\partial f_2(x_0)}{\partial x^1} & \dfrac{\partial f_2(x_0)}{\partial x^2} - f^{\#}(x_0) & \cdots & \dfrac{\partial f_2(x_0)}{\partial x^n} \\[3mm]
\vdots & \vdots & \vdots & \vdots \\[3mm]
\dfrac{\partial f_n(x_0)}{\partial x^1} & \dfrac{\partial f_n(x_0)}{\partial x^2} & \cdots & \dfrac{\partial f_n(x_0)}{\partial x^n} - f^{\#}(x_0)
\end{pmatrix}
$$

and because B is a skew-symmetric matrix, we must have the relations at 2 of the theorem.

$2 \Rightarrow 1$ Consider the Taylor expansions of $f_1, \ldots, f_n$ around x_0:

$$
f_1(x) = f_1(x_0) + \sum_{i=1}^{n} \frac{\partial f_1(x_0)}{\partial x^i}(x^i - x_0^i) + u_1(x)\|x - x_0\|
$$

$$
\vdots
$$

$$
f_n(x) = f_n(x_0) + \sum_{i=1}^{n} \frac{\partial f_n(x_0)}{\partial x^i}(x^i - x_0^i) + u_n(x)\|x - x_0\|,
$$

where $\liminf_{x \to x_0} u_i(x) = 0, \forall i \in \{1, \ldots, n\}$. Usage of the above formulae gives

$$
\frac{\langle f(x) - f(x_0), x - x_0 \rangle}{\|x - x_0\|^2} = \frac{1}{\|x - x_0\|^2}\left[\sum_{i,j=1}^{n} \frac{\partial f_i(x_0)}{\partial x^j}(x^i - x_0^i)(x^j - x_0^j) \right.
$$

$$
\left. + \sum_{i=1}^{n} u_i(x)\|x - x_0\|(x^i - x_0^i)\right].
$$

By the relations (a) and (b) one obtains

$$
\frac{\langle f(x) - f(x_0), x - x_0 \rangle}{\|x - x_0\|^2} = \frac{1}{\|x - x_0\|^2}\left[\frac{\partial f_1(x_0)}{\partial x^1}\|x - x_0\|^2\right.
$$

$$
\left. + \sum_{i=1}^{n} u_i(x)\|x - x_0\|(x^i - x_0^i)\right] = \frac{\partial f_1(x_0)}{\partial x^1} + \sum_{i=1}^{n} \frac{u_i(x)(x^i - x_0^i)}{\|x - x_0\|},
$$

wherefrom, because $-1 \le x^i - x_0^i/\|x - x_0\| \le 1$ and $\liminf_{x \to x_0} u_i(x) = 0$, it follows that

$$
f^{\#}(x_0) = \liminf_{x \to x_0} \frac{\langle f(x) - f(x_0), x - x_0 \rangle}{\|x - x_0\|^2} = \frac{\partial f_1(x_0)}{\partial x^1} = \cdots = \frac{\partial f_n(x_0)}{\partial x^n}.
$$

$\square$

1.1.3 Case $n = 2$. The Relation of the Scalar Derivative with the Complex Derivative

We identify in this chapter the complex numbers with points in $\mathbb{R}^2$. The scalar product of these numbers means the scalar product of the vectors representing them in $\mathbb{R}^2$.

THEOREM 1.11 *Let* $f : \mathbb{C} \to \mathbb{C}$ *be a complex function. The following statements are equivalent.*

1. f *is differentiable in* z_0 *as a complex function.*

2. f *is differentiable in* z_0 *as a mapping* $f : \mathbb{R}^2 \to \mathbb{R}^2$ *and is scalarly differentiable in this point.*

Proof. Follows directly from Theorem 1.10. □

The differentiability condition of f at z_0 in 2. is essential. In examples 2 and 3. of Section 1.1.4 we construct two discontinuous mappings at 0, which are scalarly differentiable in this point.

REMARK 1.3

1. *Let* G *be an open subset of* $\mathbb{C}$. *Then* f *is holomorphic on* G *if and only if it is differentiable as a vector function and scalarly differentiable on* G. *As is well known, the set of holomorphic functions on* $\mathbb{C}$ *is closed with respect to the compositions of functions.*

2. *The above remark justifies the following generalization of a holomorphic function.*

DEFINITION 1.12 *Let* G *be open in* $\mathbb{R}^n$. *The mapping* $f : \mathbb{R}^n \to \mathbb{R}^n$ *is called* $\mathbb{R}$-holomorphic *on* G *if and only if it is differentiable and scalarly differentiable on* G. *The set of* $\mathbb{R}$-holomorphic *mappings on* G *is denoted by* $\mathcal{H}(G)$.

THEOREM 1.13 *For the complex function* $f : \mathbb{C} \to \mathbb{C}$ *the following statements are equivalent*

1. f *is differentiable in* $z_0 \in \mathbb{C}$ *as a complex function.*

2. f *and* if *are scalarly differentiable in* z_0.

Proof. Let us denote $f = u + \mathrm{i}v$, $z = x + \mathrm{i}y$, $z_0 = x_0 + \mathrm{i}y_0$. Then

$$
\frac{f(z) - f(z_0)}{z - z_0} = \frac{u(z) - u(z_0) + \mathrm{i}(v(z) - v(z_0))}{x - x_0 + \mathrm{i}(y - y_0)} =
$$

$$
= \frac{[u(z) - u(z_0) + \mathrm{i}(v(z) - v(z_0))][x - x_0 - \mathrm{i}(y - y_0)]}{(x - x_0)^2 + (y - y_0)^2}
$$

$$
= \frac{[u(z) - u(z_0)](x - x_0) + [v(z) - v(z_0)](y - y_0)}{(x - x_0)^2 + (y - y_0)^2}
$$

$$
+ \mathrm{i}\frac{[v(z) - v(z_0)](x - x_0) - [u(z) - u(z_0)](y - y_0)}{(x - x_0)^2 + (y - y_0)^2}
$$

$$
= \frac{\langle f(z) - f(z_0),\, z - z_0 \rangle}{|z - z_0|^2} - \mathrm{i}\frac{\langle (\mathrm{i}f)(z) - (\mathrm{i}f)(z_0),\, z - z_0 \rangle}{|z - z_0|^2}.
$$

From the obtained relation it follows that

$$
\lim_{z \to z_0} \frac{f(z) - f(z_0)}{z - z_0}
$$

exists; that is, f is differentiable in z_0 as a complex function if and only if the limits

$$
\lim_{z \to z_0} \frac{\langle f(z) - f(z_0),\, z - z_0 \rangle}{|z - z_0|^2}
$$

and

$$
\lim_{z \to z_0} \frac{\langle (\mathrm{i}f)(z) - (\mathrm{i}f)(z_0),\, z - z_0 \rangle}{|z - z_0|^2}
$$

exist. $\qquad\qquad\square$

REMARK 1.4 *The function f is holomorphic on the open set $G \subset C$ if and only if f and $\mathrm{i}f$ are scalarly differentiable on G.*

THEOREM 1.14 *If $f : \mathbb{C} \to \mathbb{C}$ is differentiable in z_0 as a complex function then f and $\mathrm{i}f$ are scalarly differentiable in z_0 and the relation*

$$
f'(z_0) = f^{\#}(z_0) - \mathrm{i}(\mathrm{i}f)^{\#}(z_0)
$$

holds. This relation is equivalent with the relations

$$
\begin{cases}
\operatorname{Re} f'(z_0) = f^{\#}(z_0), \\[2mm]
\operatorname{Im} f'(z_0) = -(\mathrm{i}f)^{\#}(z_0).
\end{cases}
$$

Proof. It is indeed contained in the proof of Theorem 1.13. $\qquad\square$

1.1.4   Miscellanea Concerning Scalar Differentiability

Examples and counterexamples

1. Let $n \in \mathbb{N}; n > 2$. Then the set $\mathcal{H}(\mathbb{R}^n)$ of the holomorphic functions on $\mathbb{R}^n$ is not closed under compositions of functions.

   Indeed consider $A : \mathbb{R}^n \to \mathbb{R}^n$ represented by the matrix $A = (a_{ij})_{i,j} = 1, \ldots, n$, where

   $$a_{ij} = \begin{cases} 1 & \text{if } i < j, \\ 0 & \text{if } i = j, \\ -1 & \text{if } i > j. \end{cases}$$

   Obviously, A is a skew-adjoint operator. Hence A is holomorphic on $\mathbb{R}^n$. Consider $A^2 = A \circ A$ and assume that it is holomorphic. Then by Theorem 1.7 it must be of the form $A^2 = B + cI_n$ with B a skew-adjoint linear operator, c a real number, and I_n the identical map. Let us denote the matrix representing A^2 by $D = (d_{ij})_{i,j=1,\ldots,n}$, then $d_{12} + d_{21} = 0$. That is

   $$(a_{11}a_{12} + \cdots + a_{1n}a_{n2}) + (a_{21}a_{11} + \cdots + a_{2n}a_{n1}) = 0.$$

   From the definition of A it follows that $2(n - 2) = 0$ and hence $n = 2$, contradicting the hypothesis on n.

   From this example and the results in Section 1.1.3 the next assertion follows.

THEOREM 1.15 *The set of scalarly differentiable linear mappings in $\mathbb{R}^n$ is closed under composition if and only if $n \leq 2$.*

From the definition of the scalar derivative the next assertion follows easily.

LEMMA 1.16 *Let $0 = (0, \ldots, 0) \in \mathbb{R}^n$ and let $f : \mathbb{R}^n \to \mathbb{R}^n$ be a mapping having the properties:*

(a) $f(0) = 0.$

(b) $\langle f(x), x \rangle = 0, \forall x \in \mathbb{R}^n.$

Then f is scalarly differentiable in 0 and $f^{\#}(0) = 0$.

Usage of this lemma allows us to construct the following two examples of discontinuous mappings at 0, which are scalarly differentiable in this point.

2. $f : \mathbb{R}^2 \to \mathbb{R}^2$ with

$$f(x,y) = \begin{cases} \left(\dfrac{y}{x^2 + y^2}, \dfrac{-x}{x^2 + y^2} \right) & \text{if} \quad x^2 + y^2 \neq 0 \\ \\ (0,0) & \text{if} \quad x = y = 0. \end{cases}$$

3. $f : \mathbb{R}^n \to \mathbb{R}^n$ $(n \geq 2)$, with

$$f(x) = \begin{cases} \dfrac{1}{\|x\|^2} Ax & \text{if} \quad x \neq 0, \\ \\ 0 & \text{if} \quad x = 0, \end{cases}$$

where A is a nonzero, linear, skew-adjoint operator.

In fact, Example 3 generalizes Example 2. Both mappings satisfy the conditions of the above lemma and hence they are scalarly differentiable in 0. Let us show that f in 3. is not continuous at 0. Because $A \neq 0$, there exists some t in $\mathbb{R}^n$ with $At \neq 0$. Put $x = \lambda t$, $\lambda > 0$. Then the relation

$$\liminf_{\lambda \downarrow 0} \frac{A(\lambda t)}{\lambda^2 \|t\|^2} = \liminf_{\lambda \downarrow 0} \frac{1}{\lambda} \frac{At}{\|t\|^2} \neq 0$$

shows that f is not continuous at 0.

4. Example of a mapping $f : \mathbb{R}^2 \to \mathbb{R}^2$ which is continuous at 0, scalarly differentiable in this point, but not differentiable as a vector function.

Consider f given by

$$f(x,y) = \begin{cases} \left(\dfrac{xy^2}{x^2 + y^2}, \dfrac{-x^2 y}{x^2 + y^2} \right) & \text{if} \quad x^2 + y^2 \neq 0, \\ \\ (0,0) & \text{if} \quad x = y = 0. \end{cases}$$

Then f fulfills the conditions of the lemma; hence it is scalarly differentiable in 0.

The continuity of the two components of f is a standard exercise in calculus.

If f is differentiable in 0, then its components f_1 and f_2 are differentiable real-valued functions. Because

$$\frac{\partial f_1(0,0)}{\partial x} = \frac{\partial f_1(0,0)}{\partial y} = 0,$$

then $df_1(0,0) = 0$. Hence we cannot have

$$\liminf_{\substack{x \to 0 \\ y \to 0}} \frac{|f_1(x,y) - f_1(0,0) - df_1(0,0)(x,y)|}{\sqrt{x^2 + y^2}} \neq 0.$$

Taking for instance $x = y > 0$ the above limit will be $1/(2\sqrt{2})$.

5. Example of a mapping $f : \mathbb{R}^2 \to \mathbb{R}^2$ which is continuous at 0, possesses partial derivatives at 0, does not satisfy the Cauchy–Riemann conditions at 0, but is scalarly differentiable in 0.

   Take

   $$\begin{cases} \left(\dfrac{-y^3}{x^2 + y^2}, \dfrac{xy^2}{x^2 + y^2} \right) & \text{if} \quad x^2 + y^2 \neq 0, \\ \\ \qquad (0,0) & \text{if} \quad x = y = 0. \end{cases}$$

   The continuity of f at 0 can be verified passing to polar coordinates. By direct verification

   $$\frac{\partial f_1(0,0)}{\partial x} = \frac{\partial f_2(0,0)}{\partial x} = \frac{\partial f_2(0,0)}{\partial y} = 0,$$

   $$\frac{\partial f_1(0,0)}{\partial y} = -1;$$

   that is, the Cauchy–Riemann conditions do not hold at 0.

   The scalar differentiability of f at 0 follows from the fact that it satisfies the conditions of the lemma.

6. Example of a mapping $f : \mathbb{R}^2 \to \mathbb{R}^2$ which is continuous at 0, satisfies the Cauchy–Riemann conditions at 0, but is not scalarly differentiable at this point.

   Take

   $$\begin{cases} \left(\dfrac{x^2 y}{x^2 + y^2}, \dfrac{x^2 y}{x^2 + y^2} \right) & \text{if} \quad x^2 + y^2 \neq 0, \\ \\ \qquad (0,0) & \text{if} \quad x = y = 0. \end{cases}$$

   As in the above examples, the components of f are continuous at 0. Furthermore,

   $$\frac{\partial f_1(0,0)}{\partial x} = \frac{\partial f_1(0,0)}{\partial y} = \frac{\partial f_2(0,0)}{\partial x} = \frac{\partial f_2(0,0)}{\partial y} = 0,$$

   that is, the Cauchy–Riemann conditions are satisfied at 0.

Assume that f is scalarly differentiable in 0. Then the limit

$$\liminf_{\substack{x\to 0 \\ y\to 0}} \frac{f_1(x,y)x + f_2(x,y)y}{x^2+y^2} = \liminf_{\substack{x\to 0 \\ y\to 0}} \frac{x^2y(x+y)}{(x^2+y^2)^2}$$

must exist. Put $x=0$ and $y\to 0$. Then this limit will be 0. Put $x=y\neq 0$, $x\to 0$. Then this limit will be $1/2$. That is, the limit does not exist.

7. Example of a nonlinear $\mathbb{R}$-holomorphic mapping $f: \mathbb{R}^n \to \mathbb{R}^n$ for arbitrary n $(n>2)$.

By direct verification it can be seen that

$$f(x^1, x^2, \ldots, x^n) = \left((x^1)^2 - (x^2)^2 - \cdots - (x^n)^2, 2x^1x^2, \ldots, 2x^1x^n\right)$$

has scalar derivative $f^{\#}(x^1, x^2, \ldots, x^n) = 2x^1$ and satisfies the Cauchy–Riemann conditions at every point.

REMARK 1.5 *In Ahlfors [1981] it is proved that there are no nonlinear mappings of other type as that in 7. which satisfies the Cauchy–Riemann equations at every point (i.e., there are no nonlinear $\mathbb{R}$-holomorphic mappings of other type). Because $\mathcal{H}(\mathbb{R}^n)$ for $n>2$ is not closed with respect to the composition (see Example 1 and Theorem 1.15), we cannot derive other holomorphic mappings in this way.*

1.1.5 Characterization of Monotonicity by Scalar Derivatives

By using the notion of an upper (lower) scalar derivative we obtain the following assertion

THEOREM 1.17 *Let G be an open convex set in $\mathbb{R}^n$. Then the following statements are equivalent.*

1. *$f: G \to \mathbb{R}^n$ is an increasing (decreasing) mapping.*

2. *$\underline{f}^{\#}(x) \geq 0$ $(\overline{f}^{\#}(x) \leq 0)$ for each x in G.*

Proof.. The implication $1 \Rightarrow 2$ is obvious.

$2 \Rightarrow 1$ Take $\varepsilon > 0$ arbitrarily and put $g = f + \varepsilon I_n$. Then

$$g^{\#}(x) = \underline{f}^{\#}(x) + \varepsilon > 0, \ \forall x \in G.$$

Take a, b in G; $a \neq b$. For x in the line segment $[a, b]$ determined by a and b, one has by hypothesis:

$$\liminf_{y\to x} \frac{\langle g(y) - g(x), y - x\rangle}{\|y - x\|^2} > 0,$$

and hence there exists $\delta(x) > 0$ such that for any y in $I_x =]x - \delta(x)(b-a), x + \delta(x)(b-a)[\subset G$, $\langle g(y) - g(x), y - x \rangle > 0$ holds as far as $y \neq x$. Obviously,

$$[a, b] \subset \bigcup_{x \in [a,b]} I_x;$$

that is, $\{I_x : x \in [a, b]\}$ is an open cover of the compact set $[a, b]$. Hence

$$[a, b] \subset I_{y_1} \cup I_{y_2} \cup \cdots \cup I_{y_{m-1}}$$

for an appropriate set $y_1, \ldots, y_{m-1}$ of points in $]a, b[$. We can suppose that $y_1, \ldots, y_{m-1}$ are ordered from a to b. Hence $a = y_0 \in I_{y_1}$, $b = y_m \in I_{y_{m-1}}$. We can also consider that no interval I_{y_i} is contained in any other. Take $\xi_i \in I_{y_{i-1}} \cap I_{y_i} \cap]y_{i-1}, y_i[$. Then by the construction of these intervals

$$\langle g(\xi_i) - g(y_{i-1}), \xi_i - y_{i-1} \rangle > 0,$$

$$\langle g(y_i) - g(\xi_i), y_i - \xi_i \rangle > 0$$

and because ξ_i is in $]y_{i-1}, y_i[$,

$$y_i - \xi_i = \alpha(y_i - y_{i-1}),$$

$$\xi_i - y_{i-1} = \beta(y_i - y_{i-1}),$$

for appropriate positive α and β. Hence

$$\langle g(\xi_i) - g(y_{i-1}), y_i - y_{i-1} \rangle > 0,$$

$$\langle g(y_i) - g(\xi_i), y_i - y_{i-1} \rangle,$$

wherefrom

$$\langle g(y_i) - g(y_{i-1}), y_i - y_{i-1} \rangle > 0.$$

But $y_i - y_{i-1} = \lambda_i(b - a)$ for some positive λ_i, and then we must also have

$$\langle g(y_i) - g(y_{i-1}), b - a \rangle > 0.$$

By summing the above relations from $i = 1$ to $i = m$, we obtain

$$\langle g(b) - g(a), b - a \rangle > 0.$$

Rewriting this relation using the definition of g we have

$$\langle f(b) - f(a), b - a \rangle + \varepsilon \|b - a\|^2 > 0.$$

By letting $\varepsilon \to 0$ we conclude that

$$\langle f(b) - f(a), b - a \rangle \geq 0.$$

The case $\overline{f}^{\#}(x) \leq 0, \forall x \in G$ can be handled similarly. $\square$

THEOREM 1.18 *Let G be an open convex set in $\mathbb{R}^n$ and suppose that $f : G \rightarrow \mathbb{R}^n$ satisfies*

$$\underline{f}^{\#}(x) > 0 \ (\overline{f}^{\#}(x) < 0), \ \forall x \in G.$$

Then f is strictly increasing (strictly decreasing) on G.

The proof of this theorem is in fact contained in the proof of Theorem 1.17.

COROLLARY 1.19 *Let $f : \mathbb{R}^n \rightarrow \mathbb{R}^n$ be given. The following statements are equivalent.*

1. *$f^{\#}(x) = 0, \forall x \in \mathbb{R}^n$.*

2. *f is an affine mapping with a skew-adjoint linear part.*

   Proof. The implication $2 \Rightarrow 1$ is trivial.

To show that $1 \Rightarrow 2$ we apply Theorem 1.17 to conclude that

$$\langle f(y) - f(x), \ y - x \rangle = 0,$$

$\forall x, \ y \in \mathbb{R}^n$, and then by usage of Theorem 1.4 we conclude Assertion 2. $\square$

THEOREM 1.20 *Let G be convex and open in $\mathbb{R}^n$ and let $f : G \rightarrow \mathbb{R}^n$ be a Gateaux differentiable mapping on G. Then the following statements are equivalent.*

1. *f is increasing (decreasing) on G.*

2. *The Gateaux differential of f is positive (negative) semi-definite in every point of G.*

   Proof. $1 \Rightarrow 2$ Suppose that $\langle f(y) - f(x), \ y - x \rangle \geq 0 \ \forall x, \ y \in \mathbb{R}^n$. Take $y = x + \lambda t$ with $\lambda \in \mathbb{R}, \ \lambda > 0$, and $t \in \mathbb{R}^n$ arbitrarily. Then

$$\left\langle \frac{f(x + \lambda t) - f(x)}{\lambda}, t \right\rangle \geq 0$$

and letting $\lambda \rightarrow 0$ we obtain for the Gateaux differential $\delta f(x)$ that $\langle \delta f(x) t, \ t \rangle \geq 0$.

$2 \Rightarrow 1$ Suppose that the Gateaux differential is positive semi-definite at each point of G. Take a, b in $G, \ a \neq b$ and $x \in [a, b]$. Then $\langle \delta f(x) t, \ t \rangle \geq 0$ for every $t \in \mathbb{R}^n$; that is,

$$\liminf_{\lambda \downarrow 0} \left\langle \frac{f(x + \lambda t) - f(x)}{\lambda}, t \right\rangle \geq 0, \ \forall t \in \mathbb{R}^n.$$

Take $t = b - a$ and $y = x + \lambda t$, then

$$\liminf_{\substack{y \to x \\ y \in ab}} \frac{\langle f(y) - f(x), y - x \rangle}{\|y - x\|^2} \geq 0,$$

where ab denotes the line determined by a and b. By an appropriate adaptation of the method in the proof of Theorem 1.17 we can conclude

$$\langle f(b) - f(a), b - a \rangle \geq 0.$$

$\square$

THEOREM 1.21 *Let G be a convex open subset of $\mathbb{R}^n$. If $f : G \to \mathbb{R}^n$ is Gateaux differentiable on G and the Gateaux differential in each x of G is positive definite (negative definite), then f is strictly increasing (strictly decreasing) on G.*

Proof. Suppose that the Gateaux differential is positive definite. Let a, b in G, $a \neq b$. Let $x \in [a, b]$. Then $\langle \delta f(x)t, t \rangle > 0$ for every $t \in \mathbb{R}^n \backslash \{0\}$; that is,

$$\liminf_{\lambda \downarrow 0} \left\langle \frac{f(x + \lambda t) - f(x)}{\lambda}, t \right\rangle > 0, \ \forall t \in \mathbb{R}^n \backslash \{0\}.$$

Take $t = b - a$ and $y = x + \lambda t$. Then

$$\liminf_{\substack{y \to x \\ y \in ab}} \frac{\langle f(y) - f(x), y - x \rangle}{\|y - x\|^2} > 0,$$

where ab denotes the line determined by a and b. Reasoning similar to that in the proof of Theorem 1.17 yields that

$$\langle f(b) - f(a), b - a \rangle > 0.$$

$\square$

1.2 Computational Formulae for the Scalar Derivative

In this section we consider relations of the scalar derivatives with the directional derivatives and with the basic notions of spectral theory. In the two-dimensional case we exhibit a geometric connection with the Kasner circle. The obtained formulae are used for determining the monotonicity domains of operators on $\mathbb{R}^n$.

1.2.1 Scalar Derivatives and Directional Derivatives

Let $f : \mathbb{R}^n \to \mathbb{R}^n$ be given and $x, h \in \mathbb{R}^n$. If the limit

$$f'(x; h) = \liminf_{t \downarrow 0} \frac{1}{t}(f(x + th) - f(x))$$

exists with t in $\mathbb{R}$, then it is called the directional derivative of f at x in the direction h. To have a geometrical sense we suppose that $h \neq 0$, but $f'(x; 0) = 0$ can be taken obviously for each x.

The operator f is called locally Lipschitz at x, if there exist a neighbourhood V of x and a positive real number L such that for any y and z in V the inequality

$$\|f(y) - f(z)\| \leq L\|y - z\|$$

holds.

By using these notions we have the following.

THEOREM 1.22 *If the operator f is locally Lipschitz at x and the directional derivative $f'(x; h)$ exists for each h, then*

$$\underline{f}^{\#}(x) = \inf_{\|h\|=1} \langle f'(x; h), h \rangle$$

and

$$\overline{f}^{\#}(x) = \sup_{\|h\|=1} \langle f'(x; h), h \rangle.$$

Proof. We have from the definitions of the lower scalar derivative and of the directional derivative that

$$\underline{f}^{\#}(x) = \liminf_{g \to 0} \frac{\langle f(x+g) - f(x), g \rangle}{\|g\|^2}$$

$$\leq \liminf_{t \downarrow 0} \left\langle \frac{f(x+th) - f(x)}{t}, h \right\rangle$$

$$= \langle f'(x; h), h \rangle, \quad \forall h, \ \|h\| = 1. \tag{1.1}$$

Suppose that $l = \underline{f}^{\#}(x)$ and consider the sequence (h_m) with $h_m \neq 0$ and $\liminf_{m \to \infty} h_m = 0$ so as to have

$$\liminf_{m \to \infty} \frac{\langle f(x+h_m) - f(x), h_m \rangle}{\|h_m\|^2} = l.$$

Put $s_m = h_m / \|h_m\|$, $m \in \mathbb{N}$. Then by the compactness of the unit sphere in $\mathbb{R}^n$, there exists a subsequence (s_{m_k}) of s_m such that $\lim_{k \to \infty} s_{m_k} = s$, $\|s\| = 1$. Because

$$\liminf_{k \to \infty} \frac{\langle f(x+h_{m_k}) - f(x), h_{m_k} \rangle}{\|h_{m_k}\|^2} = l,$$

we have

$$l = \liminf_{k \to \infty} \left\langle \frac{f(x + \|h_{m_k}\|s) - f(x)}{\|h_{m_k}\|}, s_{m_k} \right\rangle$$

$$+ \liminf_{k \to \infty} \frac{\langle f(x + \|h_{m_k}\|s_{m_k}) - f(x + \|h_{m_k}\|s), h_{m_k} \rangle}{\|h_{m_k}\|^2},$$

wherefrom

$$l = \langle f'(x; s), s \rangle + \liminf_{k \to \infty} \frac{\langle f(y_{m_k}) - f(z_{m_k}), h_{m_k} \rangle}{\|h_{m_k}\|^2} \quad (1.2)$$

with $y_{m_k} = x + \|h_{m_k}\|s_{m_k}$, $z_{m_k} = x + \|h_{m_k}\|s$.

We have

$$\lim_{k \to \infty} y_{m_k} = \lim_{k \to \infty} z_{m_k} = x \quad (1.3)$$

and

$$\|y_{m_k} - z_{m_k}\| = \|h_{m_k}\| \, \|s_{m_k} - s\|. \quad (1.4)$$

Because f is locally Lipschitz at x, there exist a neighbourhood V of x and a real number $L > 0$ such that for any y and z in V

$$\|f(y) - f(z)\| \le L\|y - z\|. \quad (1.5)$$

From relations (1.3) there exists a k_0 such that for every $k \ge k_0$ one has y_{m_k}, $z_{m_k} \in V$. Hence by (1.4) and (1.5) it follows that

$$\|f(y_{m_k}) - f(z_{m_k})\| \le L\|h_{m_k}\| \, \|s_{m_k} - s\| \quad \forall k \ge k_0,$$

wherefrom

$$\frac{|\langle f(y_{m_k}) - f(z_{m_k}), h_{m_k} \rangle|}{\|h_{m_k}\|^2} \le \frac{\|f(y_{m_k}) - f(z_{m_k})\|}{\|h_{m_k}\|} \le L\|s_{m_k} - s\|,$$

and then

$$\liminf_{k \to \infty} \frac{\langle f(y_{m_k}) - f(z_{m_k}), h_{m_k} \rangle}{\|h_{m_k}\|^2} = 0.$$

By using now (1.0), (1.2), and the obtained relation, we conclude that

$$\underline{f^\#}(x) = \inf_{\|h\|=1} \langle f'(x; h), h \rangle.$$

The second relation in the theorem can be deduced in a similar way. $\qquad \square$

THEOREM 1.23 *If the operator f is Fréchet differentiable in x, with the differential $df(x)$, then*

$$\underline{f}^{\#}(x) = \inf_{\|h\|=1} \langle df(x)(h),\, h \rangle,$$

$$\overline{f}^{\#}(x) = \sup_{\|h\|=1} \langle df(x)(h),\, h \rangle.$$

Proof. We have by definition

$$\underline{f}^{\#}(x) = \liminf_{g \to 0} \frac{\langle f(x+g) - f(x),\, g \rangle}{\|g\|^2} \leq \liminf_{t \downarrow 0} \left\langle \frac{f(x+th) - f(x)}{t},\, h \right\rangle$$

$$= \langle df(x)(h),\, h \rangle$$

for each h, $\|h\| = 1$. Hence

$$\underline{f}^{\#}(x) \leq \inf_{\|h\|=1} \langle df(x)(h),\, h \rangle. \tag{1.6}$$

Conversely, for the expression

$$\frac{f(x+g) - f(x) - df(x)(g)}{\|g\|} = w(x, g)$$

we have $\lim\limits_{g \to 0} w(x, g) = 0$, hence

$$\underline{f}^{\#}(x) = \liminf_{g \to 0} \frac{\langle f(x+g) - f(x),\, g \rangle}{\|g\|^2}$$

$$= \liminf_{g \to 0} \left\langle w(x, g) + df(x)\left(\frac{g}{\|g\|}\right),\, \frac{g}{\|g\|} \right\rangle$$

$$= \liminf_{g \to 0} \left\langle w(x, g),\, \frac{g}{\|g\|} \right\rangle + \liminf_{g \to 0} \left\langle df(x)\left(\frac{g}{\|g\|}\right),\, \frac{g}{\|g\|} \right\rangle.$$

In the last relation we used the fact that the first limit exists inasmuch as

$$\left| \left\langle w(x, g),\, \frac{g}{\|g\|} \right\rangle \right| \leq \|w(x, g)\| \to 0 \text{ for } g \to 0.$$

Hence,

$$\underline{f}^{\#}(x) = \liminf_{g \to 0} \left\langle df(x)\left(\frac{g}{\|g\|}\right),\, \frac{g}{\|g\|} \right\rangle \geq \inf_{\|h\|=1} \langle df(x)(h),\, h \rangle.$$

Compare this relation with (1.6) to conclude that

$$\underline{f}^{\#}(x) = \inf_{\|h\|=1} \langle df(x)(h),\, h \rangle.$$

A similar method yields the proof of the second relation of the theorem. $\square$

REMARK 1.6 *Let us consider the mapping* $f : \mathbb{R}^2 \to \mathbb{R}^2$ *given by*

$$f(x, y) = \begin{cases} \left(\dfrac{xy}{\sqrt{x^2 + y^2}}, \dfrac{xy}{\sqrt{x^2 + y^2}} \right) & \text{if } x^2 + y^2 \neq 0 \\ \\ (0, 0) & \text{if } x = y = 0. \end{cases}$$

Then f *is locally Lipschitz at every* (x, y) *and has directional derivatives in each direction. Hence Theorem 1.22 applies, but not Theorem 1.23, because* f *is not differentiable in* $(0, 0)$.

By using Theorem 1.22 we get that

$$\underline{f}^{\#}(0, 0) = -\frac{\sqrt{2}}{2}, \; \overline{f}^{\#}(0, 0) = \frac{\sqrt{2}}{2}.$$

(Thanks are due to Prof. J. Kolumbán who suggested the above example to us.)

COROLLARY 1.24 *Let the operator* f *be Fréchet differentiable in* x. *Then*

1. $\underline{f}^{\#}(x) \geq 0$, $(\overline{f}^{\#}(x) \leq 0)$ *if and only if* $df(x)$ *is positive semi-definite (negative semi-definite).*

2. $\underline{f}^{\#}(x) > 0$, $(\overline{f}^{\#}(x) < 0)$ *if and only if* $df(x)$ *is positive definite (negative definite).*

3. f *is scalarly differentiable in* x *and* $f^{\#}(x) = 0$ *if and only if* $df(x)$ *is a skew-adjoint linear operator.*

Proof. The assertions 1 and 2 are obvious from Theorem 1.23. For 3 compare this theorem with Theorem 1.3. $\square$

For $A : \mathbb{R}^n \to \mathbb{R}^n$ a linear operator, we denote by A_s the operator $(A + A^*)/2$, where A^* is the adjoint of A. Let $\sigma(A_s)$ be the spectrum of A_s. With this notation we have the following.

LEMMA 1.25 *If* $A : \mathbb{R}^n \to \mathbb{R}^n$ *is a linear operator, then*

$$\min_{\|h\|=1} \langle Ah, h \rangle = \min_{\|h\|=1} \sigma(A_s), \; \max_{\|h\|=1} \langle Ah, h \rangle = \max_{\|h\|=1} \sigma(A_s).$$

Proof.

$$\min_{\|h\|=1} \langle Ah, h \rangle = \min_{\|h\|=1} \left\langle \frac{A + A^*}{2} h, h \right\rangle = \min_{\|h\|=1} \langle A_s h, h \rangle.$$

Because A_s is self-adjoint by the well-known relation we have $\min\limits_{\|h\|=1} \langle A_s h, h \rangle = \min\limits_{\|h\|=1} \sigma(A_s)$. $\hspace{2cm}$ □

THEOREM 1.26 *Let f be Fréchet differentiable in x with the differential $df(x)$. Then we have*

$$\underline{f}^{\#}(x) = \min \sigma(df(x)_s); \quad \overline{f}^{\#}(x) = \max \sigma(df(x)_s).$$

Proof. The assertion follows directly from Theorem 1.23 by using Lemma 1.25. $\hspace{7cm}$ □

COROLLARY 1.27 *Let $f : \mathbb{R}^n \to \mathbb{R}^n$ be differentiable in the convex open set G in $\mathbb{R}^n$. Consider the following assertions.*

1. *For each x in G all the eigenvalues of $df(x)_s$ are nonnegative (nonpositive).*

2. *For each x in G all the eigenvalues of $df(x)_s$ are positive (negative).*

3. *f is increasing (decreasing) on G.*

4. *f is strictly increasing (strictly decreasing) on G.*

Then $1 \Leftrightarrow 3$ and $2 \Rightarrow 4$.

Proof. Compare Theorem 1.26 with Theorems 1.17 and 1.18. $\hspace{2cm}$ □

1.2.2   Applications

A. The two-dimensional case. We use in this section the terminology and some facts from complex analysis in the form they are expressed in Hamburg et al. [1982] and Chabat [1991].

We recall that a function $f = u + iv : \mathbb{C} \to \mathbb{C}$ is called $\mathbb{R}$-*differentiable* at z in $\mathbb{C}$ if f considered as operator from $\mathbb{R}^2$ to $\mathbb{R}^2$ by the identification of $\mathbb{C}$ with $\mathbb{R}^2$ is Fréchet differentiable at z.

THEOREM 1.28 *If $f = u + iv : \mathbb{C} \to \mathbb{C}$ is $\mathbb{R}$-differentiable in $z \in \mathbb{C}$ then*

$$\underline{f}^{\#}(z) = \mathrm{Re}\frac{\partial f}{\partial z} - \left|\frac{\partial f}{\partial \overline{z}}\right|,$$

$$\overline{f}^{\#}(z) = \mathrm{Re}\frac{\partial f}{\partial z} + \left|\frac{\partial f}{\partial \overline{z}}\right|,$$

with

$$\frac{\partial f}{\partial z} = \frac{1}{2}\left(\frac{\partial f}{\partial x} - i\frac{\partial f}{\partial y}\right)$$

and

$$\frac{\partial f}{\partial \bar{z}} = \frac{1}{2} \left(\frac{\partial f}{\partial x} + i \frac{\partial f}{\partial y} \right)$$

taken in z.

Proof. Let $z = x + iy$. Then the Fréchet differential $df(z)$ of f in z can be identified with the matrix

$$\begin{bmatrix} \dfrac{\partial u}{\partial x} & \dfrac{\partial u}{\partial y} \\ \\ \dfrac{\partial v}{\partial x} & \dfrac{\partial v}{\partial y} \end{bmatrix}$$

taken in (x, y). We have then that

$$df(z)_s = \frac{df(z) + df(z)^*}{2} = \begin{bmatrix} \dfrac{\partial u}{\partial x} & \dfrac{1}{2} \left(\dfrac{\partial u}{\partial y} + \dfrac{\partial v}{\partial x} \right) \\ \\ \dfrac{1}{2} \left(\dfrac{\partial u}{\partial y} + \dfrac{\partial v}{\partial x} \right) & \dfrac{\partial v}{\partial y} \end{bmatrix}.$$

By solving the characteristic equation of this matrix we get the eigenvalues:

$$\lambda_{1,2} = \frac{1}{2} \left\{ \frac{\partial u}{\partial x} + \frac{\partial v}{\partial y} \pm \left[\left(\frac{\partial u}{\partial x} - \frac{\partial v}{\partial y} \right)^2 + \left(\frac{\partial u}{\partial y} + \frac{\partial v}{\partial x} \right)^2 \right]^{1/2} \right\}$$

and hence from Theorem 1.26

$$\underline{f}^{\#}(x) = \frac{1}{2} \left(\frac{\partial u}{\partial x} + \frac{\partial v}{\partial y} \right) - \frac{1}{2} \left[\left(\frac{\partial u}{\partial x} - \frac{\partial v}{\partial y} \right)^2 + \left(\frac{\partial u}{\partial y} + \frac{\partial v}{\partial x} \right)^2 \right]^{1/2}$$

$$\overline{f}^{\#}(x) = \frac{1}{2} \left(\frac{\partial u}{\partial x} + \frac{\partial v}{\partial y} \right) + \frac{1}{2} \left[\left(\frac{\partial u}{\partial x} - \frac{\partial v}{\partial y} \right)^2 + \left(\frac{\partial u}{\partial y} + \frac{\partial v}{\partial x} \right)^2 \right]^{1/2}$$

Now taking into account the definitions of $\partial f / \partial z$ and $\partial f / \partial \bar{z}$ we conclude the proof. $\qquad \square$

Theorem 1.28 together with Corollary 1.27 yield the following.

COROLLARY 1.29 *Let* $f : \mathbb{C} \to \mathbb{C}$ *be an* $\mathbb{R}$*-differentiable function on the convex open set G in* $\mathbb{C}$*. Then the following assertions are equivalent.*

1. *f is increasing (decreasing) on G.*

2. $\mathrm{Re} \dfrac{\partial f}{\partial z} \geq \left| \dfrac{\partial f}{\partial \bar{z}} \right|$ $\left(\mathrm{Re} \dfrac{\partial f}{\partial z} \leq - \left| \dfrac{\partial f}{\partial \bar{z}} \right| \right)$ *for every z in G.*

REMARK 1.7 *Let $f : \mathbb{C} \to \mathbb{C}$ be given and suppose that f is $\mathbb{R}$-differentiable in z. Then we have the following expression for the (complex) directional derivative in the direction $h \neq 0$ (see Hamburg et al. [1982] Proposition 2.91 or Chabat [1991] p. 34):*

$$f'(z; h) = \frac{\partial f}{\partial z} + \frac{\partial f}{\partial \bar{z}} e^{-2i\theta}$$

with $\theta = \arg h$. That is, because $\partial f / \partial z$ and $\partial f / \partial \bar{z}$ do not depend on h (they are taken as above in z), the directional derivatives describe a circle, the so-called Kasner's circle, when h varies $\partial f / \partial z$ is the centre and $|\partial f / \partial \bar{z}|$ is the radius of the circle. Hence $\underline{f}^{\#}(z)$ is the minimal oriented distance of the Kasner circle from the imaginary axis; $\overline{f}^{\#}(z)$ is the maximal oriented distance of this circle from the imaginary axis.

Thus Corollary 1.29 has the following equivalent form.

COROLLARY 1.30 *Let $f : \mathbb{C} \to \mathbb{C}$ be an $\mathbb{R}$-differentiable function on the convex open set G of $\mathbb{C}$. Then f is increasing (decreasing) on G if and only if the Kasner circle is contained for every z in G in the closed right half-plane (closed left half-plane) of the plane $\mathbb{C}$.*

Examples

1. Let $f : \mathbb{R}^2 \to \mathbb{R}^2$ be given by $f(x, y) = (xy, x + y)$. Let us identify $\mathbb{R}^2$ with $\mathbb{C}$ and consider f to be a complex function. Then

$$\frac{\partial f}{\partial z} = \frac{1}{2}\left(\frac{\partial f}{\partial x} - i\frac{\partial f}{\partial y}\right) = \frac{1}{2}[y + 1 + i(1 - x)]$$

(with $z = x + iy$) and thus $\partial f / \partial z = (y + 1)/2$.

$$\frac{\partial f}{\partial \bar{z}} = \frac{1}{2}\left(\frac{\partial f}{\partial x} + i\frac{\partial f}{\partial y}\right) = \frac{1}{2}[y - 1 + i(x + 1)]$$

and hence

$$\left|\frac{\partial f}{\partial \bar{z}}\right| = \frac{1}{2}[(x + 1)^2 + (y - 1)^2]^{1/2}.$$

We have from Theorem 1.28 that

$$\underline{f}^{\#}(x, y) = \frac{y + 1}{2} - \frac{1}{2}[(x + 1)^2 + (y - 1)^2]^{1/2},$$

$$\overline{f}^{\#}(x, y) = \frac{y + 1}{2} + \frac{1}{2}[(x + 1)^2 + (y - 1)^2]^{1/2},$$

wherefrom we get that $\underline{f}^{\#}(x, y) > 0$ is equivalent with

$$y > \left(\frac{x+1}{2}\right)^2.$$

That is, f is strictly increasing in the domain above the parabola

$$y = \left(\frac{x+1}{2}\right)^2$$

(see Theorem 1.17).

The condition $\overline{f}^{\#}(x, y) \leq 0$ yields a contradiction; hence f is nowhere decreasing (see Theorem 1.17).

2. Let $f : \mathbb{R}^2 \rightarrow \mathbb{R}^2$ be given by $f(x, y) = (\sin x - y, x + \sin y)$. The symmetrization of the Jacobi matrix of this operator is then

$$(df(x, y))_s = \begin{bmatrix} \cos x & 0 \\ 0 & \cos y \end{bmatrix}.$$

The eigenvalues of this matrix are $\cos x$ and $\cos y$ and hence we have according to Theorem 1.26 that

$$\underline{f}^{\#}(x, y) = \min\{\cos x, \cos y\},$$

$$\overline{f}^{\#}(x, y) = \max\{\cos x, \cos y\}.$$

We have $\underline{f}^{\#}(x, y) > 0$ and f is thus strictly increasing for

$$(x, y) \in \left(-\frac{\pi}{2}, \frac{\pi}{2} \right) \times \left(-\frac{\pi}{2}, \frac{\pi}{2} \right).$$

Similarly, $\overline{f}^{\#}(x, y) < 0$ and f is strictly decreasing for

$$(x, y) \in \left(\frac{\pi}{2}, \frac{3\pi}{2} \right) \times \left(\frac{\pi}{2}, \frac{3\pi}{2} \right).$$

Similar assertions hold for 2π multiple translations of the above domains in the direction of the x-axis and the y-axis.

B. The case $n > 2$
   Examples

1. Let $f : \mathbb{R}^n \to \mathbb{R}^n$ be given by

$$f(x^1, x^2, \ldots, x^n) = \left((x^1)^2 + \cdots + (x^n)^2, -2x^1 x^2, \ldots, -2x^1 x^n\right).$$

By the symmetrization of the Jacobi matrix we get

$$(df(x))_s = \begin{bmatrix} 2x^1 & 0 & \cdots & 0 \\ 0 & -2x^1 & \cdots & 0 \\ \cdots & \cdots & \cdots & \cdots \\ 0 & 0 & \cdots & -2x^1 \end{bmatrix},$$

whose eigenvalues are $\lambda_1 = 2x^1$, $\lambda_2 = \cdots = \lambda_n = -2x^1$. By Theorem 1.23 we have

$$\underline{f}^\#(x) = -2|x^1| \quad \overline{f}^\#(x) = 2|x^1|.$$

Hence $\underline{f}^\#(x) \geq 0$ or $\overline{f}^\#(x) \leq 0$ if and only if $x^1 = 0$. Thus f is nowhere strictly decreasing or strictly increasing.

2. Consider the mapping $f : \mathbb{R}^3 \to \mathbb{R}^3$ given by

$$f(x, y, z) = (x^2, y + z, z).$$

Then

$$df(x, y, z)_s = \begin{bmatrix} 2x & 0 & 0 \\ 0 & 1 & 1/2 \\ 0 & 1/2 & 1 \end{bmatrix}.$$

It can be checked that $df(x, y, z)_s$ is strictly positive definite for $x > 0$. From Theorem 1.21 it follows that f is strictly increasing for $x > 0$.

1.3   Monotonicity, Scalar Differentiability, and Conformity

In this section we relate the existence of the scalar derivative to the notion of conformity. The results of this section are based on the paper [Nemeth, 1997]. In the global case the scalar differentiability of a map (which can be identified with a vector field) on a convex open domain coincides with the conformity of the one-parameter transformation group generated by this map (vector field). In the local case the correspondence is given using the conformal derivative, a new notion introduced by us.

We also present a more geometrical interpretation of the well-known correspondence between nonexpansive (expansive) maps and decreasing (increasing) ones Zeidler [1990], which generalize the Lie correspondence between skew-adjoint maps and isometries.

1.3.1 The Coefficient of Conformity and the Conformal Derivative

DEFINITION 1.31 *Let* $f : \mathbb{R}^n \to \mathbb{R}^n$ *be a mapping and* $p \in \mathbb{R}^n$. *If the limit*

$$f^c(p) = \liminf_{q \to p} \frac{\|f(q) - f(p)\|}{\|q - p\|},$$

exists then f *will be called* conformally differentiable in p *and* $f^c(p)$ *the conformal derivative of* f *in* p. *If* f *is conformally differentiable in each point of a subset* U *of* $\mathbb{R}^n$ *then we say that* f *is* conformally differentiable on U.

LEMMA 1.32 *Let* $f : \mathbb{R}^n \to \mathbb{R}^n$ *be a differentiable mapping in* $p_0 \in \mathbb{R}^n$. *Then* f *is conformally differentiable in* p_0 *if and only if*

$$\|df_{p_0}(v)\| = \lambda_0 \|v\|$$

for all vectors v *where* λ_0 *is some nonnegative constant. If* $f : \mathbb{R}^n \to \mathbb{R}^n$ *is differentiable on* $\mathbb{R}^n$, *then* f *is conformally differentiable on* $\mathbb{R}^n$ *if and only if*

$$\|df_p(v)\| = \lambda(p)\|v\|$$

for all $p \in \mathbb{R}^n$ *and for all vectors* v, *where* $\lambda(p)$ *is some nonnegative real-valued function of* p *which does not depend on* v. *In this case*

$$f^c(p) = \lambda(p).$$

Proof. $\Longrightarrow$ Let $q = p + tv$, so that $t > 0$, $t \to 0$ and v is an arbitrary but fixed vector. Then $q \to p$, so

$$f^c(p) = \liminf_{t \to 0} \frac{\|f(p + tv) - f(p)\|}{|t|\|v\|} = \frac{1}{\|v\|}\|df_p(v)\|,$$

from where we obtain the required equality with $\lambda(p) = f^c(p)$.

$\Longleftarrow$ We have

$$\frac{\|f(p + v) - f(p)\|}{\|v\|} = \sqrt{\frac{\|f(p + v) - f(p)\|^2}{\|v\|^2}} = \sqrt{\frac{\|df_p(v) + \omega(p, v)\|v\| \|^2}{\|v\|^2}}$$

$$= \sqrt{\frac{\langle df_p(v) + \omega(p,v)\|v\|, df_p(v) + \omega(p,v)\|v\|\rangle}{\|v\|^2}}$$

$$= \sqrt{\frac{\|df_p(v)\|^2}{\|v\|^2} + 2\frac{\langle df_p(v), \omega(p,v)\rangle}{\|h\|} + \|\omega(p,v)\|^2}$$

$$= \sqrt{\lambda(p)^2 + 2\frac{\langle df_p(v), \omega(p,v)\rangle}{\|h\|} + \|\omega(p,v)\|^2}, \tag{1.7}$$

where

$$\omega(p,v) \longrightarrow 0, \quad \text{whenever } v \to 0. \tag{1.8}$$

On the other hand,

$$\frac{|\langle df_p(v), \omega(p,v)\rangle|}{\|v\|} \leq \frac{\|df_p(v)\|\|\omega(p,v)\|}{\|v\|} = \lambda(p)\|\omega(p,v)\|.$$

So

$$\frac{\langle df_p(v), \omega(p,v)\rangle}{\|v\|} \longrightarrow 0, \quad \text{whenever } v \to 0. \tag{1.9}$$

Relations (1.6), (1.8), and (1.9) imply that the limit

$$\liminf_{v \to 0} \frac{\|f(p+v) - f(p)\|}{\|v\|}$$

exists and is equal to $\lambda(p)$. So f is conformally differentiable at p and $f^c(p) = \lambda(p)$.   □

THEOREM 1.33 *Let* $f : U \to \mathbb{R}^n$ *be a smooth map, where* $U \subset \mathbf{R}^n$ *is an open set. Then* f *is conformal if and only if it is conformally differentiable on* U *and* $f^c(p) \neq 0$, *for all* $p \in U$. *In this case the coefficient of conformity* $\lambda(p)$ *is equal to* $f^c(p)$ *in each* $p \in U$.

The proof is a straightforward consequence of Lemma 1.32.

THEOREM 1.34 *Let* $f : U \to \mathbb{R}^n$ *be a smooth map where* $U \subset \mathbb{R}^n$ *is an open set. Then* f *is conformal if and only if*

$$df_p^* \circ df_p = \lambda(p)^2 I,$$

for all p *and some real valued positive function* $\lambda(p)$, *where* df_p^* *denotes the adjoint operator of* df_p *and* I *the identity operator.*

This theorem is an easy consequence of Lemma 1.32.

1.3.2   Monotone Vector Fields and Expansive Maps

The following theorem is a well-known result from the theory of Lie groups, which states that the Lie algebra of $O(n)$ is the set of skew-symmetric matrices. However, because we use the same idea of the proof of this theorem and because this is an easy proof of a classical result we state it and prove it.

THEOREM 1.35 *Let v be a vector field on $\mathbb{R}^n$ and $\psi(\varepsilon, p)$ the one parameter transformation group generated by v. Then v is skew-adjoint as a mapping from $\mathbb{R}^n$ to $\mathbb{R}^n$ if and only if $\psi(\varepsilon, p)$ is an isometry for every ε fixed.*

Proof. Theorem 1.3 implies that

$$\langle v(\psi(\varepsilon, q)) - v(\psi(\varepsilon, p)), \psi(\varepsilon, q) - \psi(\varepsilon, p) \rangle = 0,$$

for all $p, q \in \mathbb{R}^n$ and $\varepsilon \in \mathbb{R}$. Hence

$$\frac{1}{2}\frac{d}{d\varepsilon}\|\psi(\varepsilon, q) - \psi(\varepsilon, p)\|^2 = 0,$$

because

$$\|\psi(\varepsilon, q) - \psi(\varepsilon, p)\|^2 = \langle \psi(\varepsilon, q) - \psi(\varepsilon, p), \psi(\varepsilon, q) - \psi(\varepsilon, p) \rangle$$

and

$$\frac{d}{d\varepsilon}\psi(\varepsilon, p) = v(\psi(\varepsilon, p)).$$

Thus we have

$$\|\psi(\varepsilon, q) - \psi(\varepsilon, p)\| = \text{constant}$$

for p, q fixed. If we put $\varepsilon = 0$ in this relation we obtain

$$\|\psi(\varepsilon, q) - \psi(\varepsilon, p)\| = \|q - p\|$$

for all p and q, because $\psi(0, p) = p$. The converse can be proved similarly. $\square$

THEOREM 1.36 *Let v be a vector field on $\mathbb{R}^n$ and $\psi(\varepsilon, p)$ ($\varepsilon > 0$) the one-parameter transformation group generated by v. Then v is increasing (decreasing) as a mapping from $\mathbb{R}^n$ to $\mathbb{R}^n$ if and only if $\|\psi(\varepsilon, p) - \psi(\varepsilon, q)\|$ is increasing (decreasing) as a real function of ε for all p, q fixed.*

Proof. $\Longrightarrow$

$$\langle v(\psi(\varepsilon, q)) - v(\psi(\varepsilon, p)), \psi(\varepsilon, q) - \psi(\varepsilon, p) \rangle \geq 0,$$

$\forall p, q \in \mathbb{R}^n$ and $\forall \varepsilon \in \mathbb{R}^n$. Hence as before

$$\frac{1}{2}\frac{d}{d\varepsilon}\|\psi(\varepsilon, q) - \psi(\varepsilon, p)\|^2 \geq 0,$$

from where it follows that

$$\|\psi(\varepsilon, q) - \psi(\varepsilon, p)\|$$

is increasing.

$\Longleftarrow$ If

$$\|\psi(\varepsilon, p) - \psi(\varepsilon, q)\|$$

is increasing so is

$$\|\psi(\varepsilon, p) - \psi(\varepsilon, q)\|^2;$$

hence

$$\frac{d}{d\varepsilon}\|\psi(\varepsilon, p) - \psi(\varepsilon, q)\|^2 \geq 0,$$

which is equivalent to

$$\langle v(\psi(\varepsilon, q)) - v(\psi(\varepsilon, p)), \psi(\varepsilon, q) - \psi(\varepsilon, p)\rangle \geq 0,$$

$\forall p, q \in \mathbb{R}^n$, and $\forall \varepsilon \in \mathbb{R}^n$. For $\varepsilon = 0$ we have that

$$\langle v(q) - v(p), q - p\rangle \geq 0$$

for all p and q. The case v decreasing can be treated similarly. $\square$

The above theorem implies the following result, found in Zeidler [1990]:

THEOREM 1.37 *Let v be a vector field on $\mathbb{R}^n$ and $\psi(\varepsilon, p)$ the one-parameter transformation group generated by v through p. Then v is increasing (decreasing) as a mapping from $\mathbb{R}^n$ to $\mathbb{R}^n$ if and only if $\psi(\varepsilon, p)$ is an expansive (nonexpansive) mapping of p for all $\varepsilon > 0$ fixed.*

Proof. If v is increasing then it follows from Theorem 1.36 that

$$\|\psi(\varepsilon, p) - \psi(\varepsilon, q)\|$$

is increasing for all p, q fixed. Particularly, $\varepsilon > 0$ yields

$$\|\psi(\varepsilon, p) - \psi(\varepsilon, q)\| \geq \|p - q\|$$

so $\psi(\varepsilon, p)$ is an expansive mapping of p. Conversely if $\psi(\varepsilon, p)$ is expansive for all $\varepsilon > 0$ fixed then we have that

$$\|\psi(\delta - \varepsilon, \psi(\varepsilon, p)) - \psi(\delta - \varepsilon, \psi(\varepsilon, q))\| \geq \|\psi(\varepsilon, p) - \psi(\varepsilon, q)\|$$

for all $\delta > \varepsilon > 0$ and all p, q from $\mathbf{R}^n$. But

$$\psi(\delta - \varepsilon, \psi(\varepsilon, p)) = \psi(\delta, p),$$

because $\psi(\varepsilon, p)$ is a one-parameter transformation group, so

$$\|\psi(\delta, p) - \psi(\delta, q)\| \geq \|\psi(\varepsilon, p) - \psi(\varepsilon, q)\|$$

for all $\delta > \varepsilon > 0$. So

$$\|\psi(\varepsilon, p) - \psi(\varepsilon, q)\|$$

is increasing as a real function of ε for all p, q fixed. Hence by Theorem 1.36 v is increasing as a mapping from $\mathbb{R}^n$ to $\mathbb{R}^n$. $\square$

because ... ... the parameter transformation proposes:

$$\|\hat{x}(t) - \hat{x}(s)\| \geq \int_s^t \|\hat{x}'(\tau)\| d\tau = |t - s|$$

for all $t, s \in S_0$.

is increasing a contradiction, for an $p \in E$ but. Hence by Theorem 1.36, ... are increasing a mapping from $\mathbb{R}^3$. $\square$

Chapter 2

Asymptotic Derivatives and Asymptotic Scalar Derivatives

Essentially this chapter has two parts. In the first part we present the notion of the *asymptotic derivative* and some results related to this notion, and in the second part we introduce the notion of the *asymptotic scalar derivative*. The results presented in the first part are necessary for understanding the notions that are given in the second part. It seems that the notion of an asymptotic derivative was introduced by the Russian school, under the name of *asymptotic linearity*. We found this notion in M. A. Krasnoselskii's work and the reader is referred to Krasnoselskii [1964a,b] and Krasnoselskii and Zabreiko [1984]. We note that the main goal of this chapter is to present the notion of the asymptotic scalar derivative and some of its applications. This chapter may be a stimulus for new research in this subject.

2.1   Asymptotic Differentiability in Banach Spaces

Let $(E, \|\cdot\|)$ and $(F, \|\cdot\|)$ be Banach spaces. Let $\mathcal{L}(E, F)$ be the Banach space of linear continuous mappings, where the norm is $\|L\| = \sup_{\|x\|=1} \|L(x)\|$, for any $L \in \mathcal{L}(E, F)$.

DEFINITION 2.1 *We say that a nonlinear mapping* $f : E \to F$ *is asymptotically linear, if there exists* $L \in \mathcal{L}(E, F)$ *such that*

$$\lim_{\|x\| \to \infty} \frac{\|f(x) - L(x)\|}{\|x\|} = 0. \tag{2.1}$$

In this case we say that L *is an* asymptotic derivative *of* f.

PROPOSITION 2.1 *If* $f : E \to F$ *is asymptotically linear, then the mapping* $L \in \mathcal{L}(E, F)$ *that satisfies* (2.1) *is unique.*

Proof. Let f be asymptotically linear and let $L_1, L_2 \in \mathcal{L}(E, F)$ be two linear mappings such that formula (2.1) is satisfied. We have

$$\lim_{\|x\| \to \infty} \frac{\|L_1(x) - L_2(x)\|}{\|x\|} \leq \lim_{\|x\| \to \infty} \frac{\|f(x) - L_1(x)\|}{\|x\|} + \frac{\|f(x) - L_2(x)\|}{\|x\|}.$$

Then, for any $\varepsilon > 0$, there exists $r > 0$ such that for any x with $\|x\| \geq r$, we have

$$\frac{\|(L_1 - L_2)(x)\|}{\|x\|} < \varepsilon,$$

which implies

$$\left\| (L_1 - L_2) \left(\frac{x}{\|x\|} \right) \right\| < \varepsilon,$$

for any x with $\|x\| \geq r$. For any $y \in S_1 = \{x \in E : \|x\| = 1\}$ we consider $x = \rho y$ with $\rho > r$ and we have

$$[\|(L_1 - L_2)(y)\| = \left\| (L_1 - L_2) \left(\frac{x}{\|x\|} \right) \right\| < \varepsilon,$$

which implies $\|L_1 - L_2\| < \varepsilon$ and finally $\|L_1 - L_2\| = 0$; that is, $L_1 = L_2$. $\square$

REMARK 2.1 *If $f : E \to F$ is asymptotically linear, then in this case we say that the linear continuous mapping L used in Definition 2.1 is the asymptotic derivative of f and we denote $L = f_\infty$.*

The following result due to M. A. Krasnoselskii is important in the study of bifurcation problems [Amann, 1973, 1974b, 1976; Krasnoselskii, 1964a,b; Krasnoselskii and Zabreiko, 1984]. We recall that a mapping $f : E \to F$ is *completely continuous*, if it is continuous, and for any bounded set $D \subset E$, we have that $f(D)$ is relatively compact.

THEOREM 2.2 *Let $f : E \to F$ be a nonlinear mapping. If f is completely continuous and asymptotically linear, then f_∞ is completely continuous.*

Proof. We use the fact that in a Banach space, a sequence is convergent if and only if it is a Cauchy sequence.

Indeed, we assume that f_∞ is not completely continuous. Then, we can define a sequence $\{x_n\}_{n \in \mathbb{N}} \subset S(0, 1) = \{x \in E : \|x\| = 1\}$ such that $\|f_\infty(x_n) - f_\infty(x_m)\| \geq 3\delta > 0$, for any n and m such that $n \neq m$. Considering formula (2.1) we deduce the existence of a real number $r > 0$ such that $\|f(x) - f_\infty(x)\| < \delta\|x\|$, for any x with $\|x\| = r$. Then, we have

$$\|f(rx_n) - f(rx_m)\| \geq \|f_\infty(rx_n) - f_\infty(rx_m)\| - \|f(rx_n) - f_\infty(rx_n)\|$$
$$- \|f_\infty(rx_m) - f(rx_m)\| > r\|f_\infty(x_n) - f_\infty(x_m)\| - 2\delta r,$$

which implies $\|f(rx_n) - f(rx_m)\| \geq \delta r$ for $n \neq m$ and the compactness is contradicted. ☐

If a nonlinear mapping has an asymptotic derivative, the computation of this derivative, generally cannot be so simple. We give now some examples.

(A) Let $G \subset \mathbb{R}^n$ be the closure of a bounded open set whose boundary is a null set (i.e., G has a piecewise smooth boundary). Consider the following Hammerstein mapping

$$A(\varphi)(t) = \int_G K(t,s) f[s, \varphi(s)] \, d s,$$

where $f : G \times \mathbb{R} \to \mathbb{R}$ and $K : G \times G \to \mathbb{R}$. Suppose that the following conditions are satisfied.

(i) $\int_G \int_G K^2(t,s) \, d t \, d s < \infty.$
(ii) The mapping $f_0(\varphi)(s) = f[s, \varphi(s)]$, $\varphi \in L^2$ is such that $f_0 : L^2 \to L^2$.
(iii) $|f(t,u) - u| \leq \sum_{j=1}^n S_j(t)|u|^{1-p_j} + D(t)$, where $t \in G$; $-\infty < u < +\infty$; $S_j(t) \in L^{2/p_j}$, $0 < p_j < 1$, $j = 1, 2, \ldots, n$, and $D(t) \in L^2$.

Consider the linear mapping

$$B(\varphi)(t) = \int_G K(t,s) \varphi(s) \, d s.$$

In this case we have that $A : L^2 \to L^2$ and $B \in \mathcal{L}(L^2, L^2)$. Because

$$\frac{\|A(\varphi) - B(\varphi)\|}{\|\varphi\|} = \frac{1}{\|\varphi\|} \left\{ \int_G \left[\int_G K(t,s)[f[s, \varphi(s)] - \varphi(s)] \, d s \right]^2 d t \right\}^{1/2}$$

$$\leq \frac{\{\int_G \int_G K^2(t,s) \, d t \, d s\}^{1/2}}{\|\varphi\|} \left\{ \sum_{j=1}^n \left[\int_G S_j^{2/p_j}(t) \, d t \right]^{p_j/2} \|\varphi\|^{1-p_j} \right.$$

$$\left. + \left[\int_G D^2(t) \, d t \right]^{1/2} \right\}$$

we deduce that

$$\lim_{\|\varphi\| \to \infty} \frac{\|A(\varphi) - B(\varphi)\|}{\|\varphi\|} = 0;$$

that is, $A_\infty = B$.

(B) Suppose that $(E, \| \cdot \|)$ and $(F, \| \cdot \|)$ are two particular Banach spaces of functions defined on a particular subset (which can be as in Example (A)).

Consider a function $f(t, u)$ where $t \in G$ and $-\infty < u < +\infty$. Generally we suppose that f satisfies *Carathéodory conditions*; that is, f is continuous with respect to u and measurable with respect to t. The operator $f^*(x)(t) = f[t, x(t)]$ is called the substitution mapping. Suppose that $f^* : E \to F$. If

$$\lim_{t \to \infty} \frac{f(t, u)}{u}$$

exists, we denote it $g(t)$. In this case the asymptotic derivative of f^* must necessarily be of the form $f_\infty(h)(t) = g(t)h(t)$. In the next section we present another interesting case when we can compute the asymptotic derivative of a nonlinear mapping (when this derivative exists).

2.2 Hyers–Ulam Stability and Asymptotic Derivatives

The Hyers–Ulam stability of functional equations offers us the possibility to compute the asymptotic derivative of an asymptotic differentiable mapping. The notion of Hyers–Ulam stability of mappings has its origin in a problem defined by S. Ulam during a talk presented in 1940, at the mathematics club of the University of Wisconsin, in which he discussed a number of unsolved problems. This problem is related to the stability of homomorphisms. Given a group G_1, a metric group G_2 with metric $d(\cdot, \cdot)$, and a real number $\varepsilon > 0$, does there exist a $\delta > 0$ such that if $f : G_1 \to G_2$ satisfies $d(f(xy), f(x)f(y)) < \delta$ for all $x, y \in G_1$, then a homeomorphism $h : G_1 \to G_2$ exists with $d(f(x), h(x)) < \varepsilon$ for all $x \in G_1$? In 1941, D. H. Hyers gave a positive answer to Ulam's problem for approximately additive mappings [Hyers, 1941]. He proved the following result. If $(E, \| \cdot \|)$ and $(F, \| \cdot \|)$ are Banach spaces and $f : E \to F$ is a mapping satisfying the condition

$$\|f(x + y) - f(x) - f(y)\| < \varepsilon$$

for all $x, y \in E$, then there is a unique additive mapping T satisfying

$$\|f(x) - T(x)\| \leq \varepsilon.$$

Thus, the stability theory of Hyers and Ulam started. We note that for almost three decades almost no progress was made on this problem, probably because the theory of functional equations was not sufficiently developed at that time. The Hyers–Ulam stability began to expand at the end of the seventies and now there is extensive literature in the subject which forms the so-called Hyers–Ulam stability theory. About this theory the reader is referred to the books by Czerwik [1994, 2001], Hyers et al. [1998a] and Rassias [1978]. We present some results from Hyers–Ulam stability theory related to the asymptotic differentiability. In 1978, a generalized solution to Ulam's problem for approximately linear mappings was given by Th. M. Rassias (see [Rassias, 1978]). Let $(E, \| \cdot \|)$

and $(F, \|\cdot\|)$ be Banach spaces. He considered a mapping $f : E \to F$ satisfying the condition of continuity of $f(tx)$ in t for each fixed x and such that $\|f(x + y) - f(x) - f(y)\| \leq \theta(\|x\|^p + \|y\|^p)$, for any $x, y \in E$ and that $T : E \to F$ is the unique linear mapping satisfying

$$\|f(x) - T(x)\| \leq \frac{2\theta}{2 - 2^p} \|x\|^p.$$

This result is valid also when $p < 0$ and when $p > 1$. The following definition is due to G. Isac.

DEFINITION 2.3 *We say that a mapping $f : E \to F$ is ψ-additive if and only if there exist $\theta > 0$ and a function $\psi : \mathbb{R}_+ \to \mathbb{R}_+$ such that*

$$\lim_{t \to \infty} \frac{\psi(t)}{t} = 0$$

and

$$f(x + y) - f(x) - f(y) \leq \theta[\psi(\|x\|) + \psi(\|y\|)]$$

for all $x, y \in E$.

In 1991 Isac and Rassias proved the following result published in Isac and Rassias [1993a].

THEOREM 2.4 *Let $(E, \|\cdot\|)$ and $(F, \|\cdot\|)$ be Banach spaces and $f : E \to F$ a mapping such that $f(tx)$ is continuous in t for each fixed x. If f is ψ-additive and ψ satisfies*

1. $\psi(ts) \leq \psi(t)\psi(s)$, *for all $t, s \in \mathbb{R}_+$;*

2. $\psi(t) < t$, *for all $t > 1$;*

then there exists a unique linear mapping $T : E \to F$ such that

$$\|f(x) - T(x)\| \leq \left[\frac{2\theta}{2 - \psi(2)}\right] \psi(\|x\|),$$

for all $x \in E_1$.

Proof. We show that

$$\left\|\frac{f(2^n x)}{2^n} - f(x)\right\| \leq \left\{\theta \sum_{m=0}^{n-1} \left[\frac{\psi(2)}{2}\right]^m\right\} \psi(\|x\|) \qquad (2.2)$$

for any positive integer n, and for any $x \in E$. The proof of (2.2) follows by induction on n. For $n = 1$ by ψ-additivity of f we have

$$\|f(2x) - 2f(x)\| \leq 2\theta\psi(\|x\|),$$

which implies

$$\left\|\frac{f(2x)}{2} - f(x)\right\| \le \theta\psi(\|x\|).$$

Assume now that (2.2) holds for n and we want to prove it for the case $n+1$. Replacing x by $2x$ in (2.2) we obtain

$$\left\|\frac{f(2^n 2x)}{2^n} - f(2x)\right\| \le \left\{\theta \sum_{m=0}^{n-1} \left[\frac{\psi(2)}{2}\right]^m\right\} \psi(2\|x\|).$$

Because $\psi(2\|x\|) \le \psi(2)\psi(\|x\|)$ we get

$$\left\|\frac{f(2^{n+1}x)}{2^n} - f(2x)\right\| \le \left\{\theta \sum_{m=0}^{n-1} \left[\frac{\psi(2)}{2}\right]^m\right\} \psi(2)\psi(\|x\|). \qquad (2.3)$$

Multiplying both sides of (2.3) by $1/2$ we obtain

$$\left\|\frac{f(2^{n+1}x)}{2^{n+1}} - \frac{f(2x)}{2}\right\| \le \left\{\theta \sum_{m=1}^{n} \left[\frac{\psi(2)}{2}\right]^m\right\} \psi(\|x\|).$$

Now, using the triangle inequality we deduce

$$\left\|\frac{1}{2^{n+1}}[f(2^{n+1}x)] - f(x)\right\| \le \left\|\frac{1}{2^{n+1}}[f(2^{n+1}x)] - \frac{1}{2}[f(2x)]\right\|$$

$$+ \left\|\frac{1}{2}[f(2x)] - f(x)\right\|$$

$$\le \left\{\theta \sum_{m=1}^{n} \left[\frac{\psi(2)}{2}\right]^m\right\} \psi(\|x\|) + \theta\psi(\|x\|)$$

$$= \theta\psi(\|x\|)\left\{1 + \sum_{m=1}^{n} \left[\frac{\psi(2)}{2}\right]^m\right\},$$

which proves (2.3). Thus,

$$\left\|\frac{1}{2^{n+1}}[f(2^{n+1}x)] - f(x)\right\| \le \theta\psi(\|x\|)\left\{1 + \sum_{m=1}^{n} \left[\frac{\psi(2)}{2}\right]^m\right\} \le \frac{2\theta\psi(\|x\|)}{2 - \psi(2)}.$$

For $m > n > 0$ we have

$$\left\|\frac{1}{2^m}[f(2^m x)] - \frac{1}{2^n}[f(2^n x)]\right\| = \frac{1}{2^n}\left\|\frac{1}{2^{m-n}}[f(2^m x)] - f(2^n x)\right\|$$

$$= \frac{1}{2^n}\left\|\frac{1}{2^r}[f(2^r y)] - f(y)\right\|,$$

where $r = m - n$ and $y = 2^n x$.

$$\left\| \frac{1}{2^m} \left[f(2^m x) \right] - \frac{1}{2^n} \left[f(2^n x) \right] \right\| \leq \frac{1}{2^n} \theta \left[\frac{2\psi(\|y\|)}{2 - \psi(2)} \right]$$

$$= \frac{1}{2^n} \theta \left[\frac{2\psi(\|y\|)}{2 - \psi(2)} \right]$$

$$\leq \frac{1}{2^n} \theta \left[\frac{2\psi(2^n)\psi(\|x\|)}{2 - \psi(2)} \right]$$

$$\leq \left[\frac{\psi(2)}{2} \right]^n \theta \left[\frac{2\psi(\|x\|)}{2 - \psi(2)} \right].$$

But because

$$\lim_{n \to \infty} \left[\frac{\psi(2)}{2} \right]^n = 0,$$

we have that

$$\left\{ \left(\frac{1}{2^n} \right) \left[f(2^n x) \right] \right\}_{n \in \mathbb{N}}$$

is a Cauchy sequence. Set

$$T(x) = \lim_{n \to \infty} \frac{f(2^n x)}{2^n},$$

for all $x \in E$. The mapping $x \mapsto T(x)$ is additive. Indeed, we have

$$\| f[2^n(x + y)] - f(2^n x) - f(2^n y) \| \leq \theta[\psi(\|2^n x\|) + \psi(\|2^n y\|)]$$

$$= \theta[\psi(2^n\|nx\|) + \psi(2^n\|y\|)]$$

$$\leq \theta\psi(2^n)[\psi(\|x\|) + \psi(\|y\|)],$$

which implies that

$$\left(\frac{1}{2^n} \right) \| f[2^n(x + y)] - f(2^n x) - f(2^n y) \|$$

$$\leq \left[\frac{\psi(2^n)}{2^n} \right] \theta[\psi(\|x\|) + \psi(\|y\|)] \leq \left[\frac{\psi(2)}{2} \right]^n \theta[\psi(\|x\|) + \psi(\|y\|)].$$

However,

$$\lim_{n \to \infty} \left[\frac{\psi(2)}{2} \right]^n = 0,$$

thus

$$\lim_{n \to \infty} \frac{1}{2^n} \| f[2^n(x + y)] - f(2^n x) - f(2^n y) \| = 0.$$

Therefore,

$$T(x+y) = T(x) + T(y), \tag{2.4}$$

for all $x, y \in E$. Because of (2.4) it follows that $T(rx) = rT(x)$ for any rational number r, which implies that $T(ax) = aT(x)$ for any real value of a. Hence, T is a linear mapping. From

$$\left\| \frac{f(2^n x)}{2^n} - f(x) \right\| \le 2\theta \frac{\psi(\|x\|)}{2 - \psi(2)},$$

taking the limit as $n \to \infty$ we obtain

$$\|T(x) - f(x)\| \le \frac{2\theta\psi(\|x\|)}{2 - \psi(2)}. \tag{2.5}$$

We claim that T is the unique such linear mapping. Suppose that there exists another one, denoted $g : E \to F$ satisfying

$$\|f(x) - g(x)\| \le \frac{2\theta_1 \psi_1(\|x\|)}{2 - \psi_1(2)}. \tag{2.6}$$

From (2.5) and (2.6) we get

$$\begin{aligned}
\|T(x) - g(x)\| &\le \|T(x) - f(x)\| + \|f(x) - g(x)\| \\
&\le \frac{2\theta\psi(\|x\|)}{2 - \psi(2)} + \frac{2\theta_1\psi_1(\|x\|)}{2 - \psi_1(2)}.
\end{aligned}$$

Then,

$$\begin{aligned}
\|T(x) - g(x)\| &= \left\| \frac{1}{n} T(nx) - \frac{1}{n} g(nx) \right\| \\
&\le \left[\frac{\psi(n)}{n} \right] \left[\frac{2\theta\psi(\|x\|)}{2 - \psi(2)} \right] + \left[\frac{\psi_1(n)}{n} \right] \left[\frac{2\theta_1\psi_1(\|x\|)}{2 - \psi_1(2)} \right],
\end{aligned}$$

for every positive integer $n > 1$. However,

$$\lim_{n \to \infty} \frac{\psi(n)}{n} = 0 = \lim_{n \to \infty} \frac{\psi_1(n)}{n}.$$

Therefore, $T(x) = g(x)$ for all $x \in E$.   □

The mapping T defined by Theorem 2.4 has some remarkable properties.

(A) If $f(S)$ is bounded, where $S = \{x \in E : \|x\| = 1\}$, in particular if f is completely continuous, then T is continuous. Indeed, this is the consequence of the inequalities

$$\|T(x)\| \quad \leq \quad \|f(x)\| + \|T(x) - f(x)\|$$

$$\leq \quad \|f(x)\| + \frac{2\theta}{2 - \psi(2)}\psi(\|x\|)$$

$$\leq \quad \|f(x)\| + \frac{2\theta}{2 - \psi(2)}\psi(1),$$

for all $x \in S$.

(B) When, the linear mapping T defined by Theorem 2.4 is continuous, in particular when $f(S)$ is bounded or f is completely continuous, we have that f is asymptotically linear and $f_\infty = T$. Indeed, we have

$$\lim_{x \to +\infty} \frac{\|f(x) - T(x)\|}{\|x\|} \leq \frac{2\theta}{2 - \psi(2)} \lim_{x \to +\infty} \frac{\psi(\|x\|)}{\|x\|} = 0.$$

The class of functions $\psi : \mathbb{R}_+ \to \mathbb{R}_+$, which satisfies conditions asked in Theorem 2.4; that is,

(i_0) $\displaystyle\lim_{t \to +\infty} \frac{\psi(t)}{t} = 0$;

(i_1) $\psi(ts) \leq \psi(t)\psi(s)$, for all $t, s \in \mathbb{R}_+$;

(i_2) $\psi(t) < t$, for all $t > 1$;

is not empty. In this case we can cite the following functions.

(1) $\psi(t) = t^p$, with $p \in [0, 1[$;

(2) $\psi(t) = \begin{cases} 0 & \text{if } t = 0, \\ t^p & \text{if } t > 0, \end{cases}$ where $p < 0$.

Now, we show that it is possible to enlarge the class of functions ψ such that the conclusion of Theorem 2.4 remains valid. Let $\mathcal{F}(\psi)$ be the set of all functions $\psi : \mathbb{R}_+ \to \mathbb{R}_+$ satisfying conditions (i_0), (i_1), and (i_2). Let $\mathcal{P}(\psi)$ be the convex cone (for the definition of a convex cone see the first section of Chapter 4) generated by the set $\mathcal{F}(\psi)$ (i.e., the smallest convex cone containing this set). We remark that a function $\psi \in \mathcal{P}(\psi)$ satisfies the assumption (i_0) but generally does not satisfy the assumptions (i_1) and (i_2). However, we show that Theorem 2.4 remains valid for ψ-additive functions with $\psi \in \mathcal{P}(\psi)$. The following result is a consequence of the main result proved in Gavruta [1994].

LEMMA 2.5 *If* $\phi : E \times E \to [0, +\infty[$ *is a mapping such that*

$$\phi_0(x, y) = \sum_{k=0}^{\infty} 2^{-k} \psi(2^k x, 2^k y) < +\infty,$$

for all $x, y \in E$ and $f : E \to F$ is a continuous mapping such that

$$\|f(x + y) - f(x) - f(y)\| \leq \phi(x, y),$$

for all $x, y \in E$, then there exists a unique linear mapping $T : E \to F$ such that

$$\|f(x) - T(x)\| \leq \frac{1}{2}\phi_0(x, x),$$

for all $x \in E$. Moreover

$$T(x) = \lim_{n \to \infty} \frac{f(2^n x)}{2^n},$$

for all $x \in E$.

A consequence of Lemma 2.5 is the following result which is a generalization of Theorem 2.4.

THEOREM 2.6 *Let $f : E \to F$ be a continuous mapping and $\psi \in \mathcal{P}(\psi)$, such that*

$$\psi = \sum_{i=1}^{m} a_i \psi_i,$$

where for each i, $a_i > 0$ and $\psi_i \in \mathcal{F}(\psi)$. If f is ψ-additive, then there exists a unique linear mapping $T : E \to F$ such that

$$\|f(x) - T(x)\| \leq 2\theta M \psi(\|x\|),$$

for any $x \in E$, where

$$M = \max\left\{ \frac{1}{2 - \psi_i(2)} : i = 1, 2, \cdots, m \right\}$$

and

$$T(x) = \lim_{n \to \infty} \frac{f(2^n x)}{2^n},$$

for any $x \in E$. Moreover,

$$\lim_{\|x\| \to \infty} \frac{\psi(\|x\|)}{\|x\|} = 0.$$

Proof. We consider the function

$$\Phi(x, y) = \theta[\psi(\|x\|) + \psi(\|y\|)],$$

for any $x, y \in E$, where θ is the constant used in the ψ-additivity assumption, and we apply Lemma 2.5. To do this first we must show that

$$\Phi_0(x, y) = \sum_{k=0}^{\infty} 2^{-k} \Phi(2^k x, 2^k y)$$

is convergent for any $x, y \in E$. Indeed, we have

$$
\Phi_0(x, y) = \theta \sum_{k=0}^{\infty} 2^{-k} \left[\sum_{i=1}^{m} a_i \psi_i(2^k \|x\|) + \sum_{i=1}^{m} a_i \psi_i(2^k \|y\|) \right]
$$

$$
= \theta \left[\sum_{i=1}^{m} a_i \sum_{k=0}^{\infty} 2^{-k} \psi_i(2^k \|x\|) + \sum_{i=1}^{m} a_i \sum_{k=0}^{\infty} 2^{-k} \psi_i(2^k \|y\|) \right]
$$

$$
\leq \theta \left\{ \sum_{i=1}^{m} a_i \sum_{k=0}^{\infty} \left[\frac{\psi_i(2)}{2} \right]^k \psi_i(\|x\|) \right.
$$

$$
\left. + \sum_{i=1}^{m} a_i \sum_{k=0}^{\infty} \left[\frac{\psi_i(2)}{2} \right]^k \psi_i(\|y\|) \right\} < \infty,
$$

because the series

$$
\sum_{k=0}^{\infty} \left[\frac{\psi_i(2)}{2} \right]^k \psi_i(\|x\|)
$$

and

$$
\sum_{k=0}^{\infty} \left[\frac{\psi_i(2)}{2} \right]^k \psi_i(\|y\|)
$$

are convergent. Applying Lemma 2.5 we have that T is well defined by

$$
T(x) = \lim_{n \to \infty} \frac{f(2^n x)}{2^n},
$$

for any $x \in E$. Because f is continuous, we have that T is not only additive as in Gavruta [1994] but it is linear too. We have

$$
f(x) - T(x) \leq \frac{1}{2} \Phi_0(x, x),
$$

for any $x \in E$. Now, we evaluate $\Phi_0(x, x)$. We have

$$
\Phi_0(x, x) = 2\theta \left\{ \sum_{i=1}^{m} a_i \sum_{k=0}^{\infty} \left[\frac{\psi_i(2)}{2} \right]^k \psi_i(\|x\|) \right\}
$$

$$
= 4\theta \sum_{i=1}^{m} a_i \psi_i(\|x\|) \frac{1}{2 - \psi_i(2)},
$$

which implies

$$\|f(x) - T(x)\| \leq 2\theta \sum_{i=1}^{m} a_i \psi_i(\|x\|) \frac{1}{2 - \psi_i(2)}$$

$$\leq 2\theta M \sum_{i=1}^{m} a_i \psi_i(\|x\|) = 2\theta M \psi(\|x\|),$$

where

$$M = \max \left\{ \frac{1}{2 - \psi_i(2)} : i = 1, 2, \ldots, m \right\}.$$

Because for any $i = 1, 2, \ldots, m$, ψ_i satisfies assumption (i_0), we have that

$$\lim_{\|x\| \to \infty} \frac{\psi_i(\|x\|)}{\|x\|} = 0,$$

which implies that

$$\lim_{\|x\| \to \infty} \frac{\psi_i(\|x\|)}{\|x\|} = 0. \qquad \square$$

REMARK 2.2

1. *If $f : E \to F$ is continuous, ψ-additive with $\psi \in \mathcal{P}(\psi)$ and $f(S)$ is bounded, then in this case we also have that $f_\infty = T$.*

2. *Theorem 2.6 is significant, because the class of ψ-additive mappings with $\psi \in \mathcal{P}(\psi)$ is strictly larger than the class of mappings defined in Theorem 2.4. In this sense we remark the following results.*

   (a) *If $f : E \to F$ is a ψ-additive mapping with $\psi \in \mathcal{P}(\psi)$ and $L \in \mathcal{L}(E, F)$, then $L + f$ is a ψ-additive mapping with respect to the same function ψ.*

   (b) *If $f : E_1 \to E_2$ is a ψ-additive mapping with $\psi \in \mathcal{P}(\psi)$ and $L \in \mathcal{L}(E_2, E_3)$, then $L \circ f$ is a ψ-additive mapping from E_1 into E_3 with respect to the same function ψ and the constant θ replaced by $\theta \|L\|$. We note that E_1, E_2, and E_3 are Banach spaces.*

   (c) *If $f_1, f_2 : E \to E$ are mappings such that f_1 is ψ_1 additive and f_2 is ψ_2 additive, then for every $a_1, a_2 \in \mathbb{R}_+ \backslash \{0\}$, we have that $a_1 f_1 + a_2 f_2$ is a ψ-additive mapping where $\psi = \psi_1 + \psi_2$ and $\theta = \max\{a_1 \theta_1, a_2 \theta_2\}$.*

More results about ψ-additivity and its generalizations are given in Czerwik [1994, 2001], Gavruta [1994], Hyers et al. [1998a,b], Isac and Rassias [1993a,b, 1994]. It is interesting to note that Theorem 2.4 was recently proved again by

V. Radu using the fixed point theory [Radu, 2003]. Perhaps Radu's method will open a new research direction in the Hyers–Ulam stability of mappings.

REMARK 2.3 *We note that the constant*

$$\frac{2\theta}{2 - \psi(2)}$$

used in Theorem 2 given in Isac and Rassias [1993b] must be the constant $2\theta M$ computed in Theorem 2.6, or the constant

$$\frac{2\theta}{2 - \psi(2)}$$

must be

$$\frac{2\theta}{2 - \psi_{i_0}(2)},$$

where

$$\frac{1}{2 - \psi_{i_0}} = M = \max \left\{ \frac{1}{2 - \psi_i(2)} \right\}_{i=1}^{m}.$$

We remarked above that under the assumptions of Theorem 2.6, if f is continuous and f(S) is bounded, we have that

$$f_\infty(x) = \lim_{n \to \infty} \frac{f(2^n x)}{2^n},$$

for any $x \in E$. Conversely, if f has an asymptotic derivative, namely,

$$f_\infty = T \in \mathcal{L}(E, F),$$

that is, if

$$\lim_{n \to \infty} \frac{\|f(x) - T(x)\|}{\|x\|},$$

then at any point $x \in E$, we have that

$$f(x) = \lim_{n \to \infty} \frac{f(2^n x)}{2^n},$$

for any $x \in E$. Indeed, if $x \in E \backslash \{0\}$, then we have that $\|2^n x\| \to \infty$ as $n \to \infty$ and

$$0 = \lim_{n \to \infty} \frac{\|f(2^n x) - T(2^n x)\|}{\|2^n x\|} = \lim_{n \to \infty} \frac{\|f(2^n x) - T(2^n x)\|}{2^n \|x\|}$$

$$= \frac{1}{\|x\|} \lim_{n \to \infty} \left\| \frac{2^n x}{2^n} - T(x) \right\|.$$

Therefore,

$$T(x) = \lim_{n \to \infty} \frac{f(2^n x)}{2^n},$$

for any $x \in E$, because this formula is also true for $x = 0$.

We recall the following classical result due to Krasnoselskii.

THEOREM 2.7 *Let $f : E \to E$ be a mapping such that for a $\rho > 0$ sufficiently large, the mapping f has a Frechét derivative denoted by $f'(x)$, at any element x with $\|x\| > \rho$, and*

$$\lim_{\|x\| \to \infty} \|f'(x) - T\| = 0,$$

where $T \in \mathcal{L}(E, F)$, then $f_\infty = T$.

Proof. This result is a particular case of Theorem 3.3 proved in the Russian edition of Krasnoselskii [1964a]. $\qquad\square$

The following result is inspired by Theorem 2.7. The Frechét derivative is replaced by a ψ-additive mapping.

THEOREM 2.8 *Let $f : E \to F$ be a mapping, $g : E \to F$ a continuous mapping such that $g(S)$ is bounded, and ψ_1 a mapping which satisfies condition (i_0). If the following two assumptions are satisfied,*

1. *there exist two constants, $\rho > 0$ and $M_1 > 0$ such that*

$$\|f(x) - g(x)\| \le M_1 \psi_1(\|x\|),$$

   for any $x \in E$ with $\|x\| > \rho$,

2. *g is ψ_2-additive, with $\psi_2 \in \mathcal{P}(\psi)$,*

then f is an asymptotically linear mapping and

$$f_\infty(x) = \lim_{n \to \infty} \frac{g(2^n x)}{2^n} = \lim_{n \to \infty} \frac{f(2^n x)}{2^n}$$

for any $x \in E$.

Proof. Because g is ψ_2-additive with $\psi_2 \in \mathcal{P}(\psi)$, we apply Theorem 2.7 and we obtain a constant $M_2 > 0$ and a continuous linear mapping $T : E \to F$ such that

$$\|g(x) - T(x)\| \le 2\theta M_2 \psi_2(\|x\|),$$

for every $x \in E$. We know that

$$T(x) = \lim_{n \to \infty} \frac{g(2^n x)}{2^n},$$

for any $x \in E$. We have

$$
\begin{aligned}
\|f(x) - T(x)\| &\leq \|f(x) - g(x)\| + \|g(x) - T(x)\| \\
&\leq M_1\psi_1(\|x\|) + 2\theta M_2\psi_2(\|x\|),
\end{aligned}
$$

for all $x \in E$, which implies

$$
\frac{\|f(x) - T(x)\|}{\|x\|} \leq \lim_{\|x\| \to \infty} M_1\psi_1(\|x\|)\|x\| + \lim_{\|x\| \to \infty} \frac{2\theta M_2\psi_2(\|x\|)}{\|x\|} = 0.
$$

$\square$

REMARK 2.4 *Krasnoselskii in [1964a] considers the following definition of the asymptotic derivative of a mapping $f : E \to F$. We say that a linear mapping $f_\infty \in \mathcal{L}(E, F)$ is the* asymptotic derivative *of f if*

$$
\lim_{n \to \infty} \sup_{\|x\| \geq R} \frac{\|f(x) - f_\infty(x)\|}{\|x\|} = 0.
$$

Many interesting applications of this notion are given in Krasnoselskii in [1964a].

2.3 Asymptotic Differentiability Along a Convex Cone in a Banach Space

Let $(E, \|\cdot\|)$ be a Banach space and $K \subset E$ a closed pointed convex cone (for a definition see the first section of Chapter 4). We say that K is a generating cone if $E = K - K$. If K has a nonempty interior then K is generating. Indeed, let $\mathring{K}$ be the interior of K. Let $v_0 \in \mathring{K}$ be an arbitrary element. For any $x \in E$, there exists $\lambda \in\,]0, 1[$ such that

$$
y = \lambda v_0 + (1 - \lambda)x \in K.
$$

If

$$
\rho = \frac{1 - \lambda}{\lambda}
$$

and

$$
z = \frac{1}{\lambda}y,
$$

then we have

$$
z = v_0 + \rho x \in K,
$$

which implies

$$
x = u - v,
$$

where

$$
u = \frac{1}{\rho}z \in K
$$

and

$$v = \frac{1}{\rho}v_0 \in K.$$

Let $(F, \| \cdot \|)$ be another Banach space and $f : K \to F$ a mapping.

DEFINITION 2.9 *We say that f is* asymptotically linear along the cone K *if there exists $T \in \mathcal{L}(E, F)$ such that*

$$\lim_{\substack{\|x\| \to \infty \\ x \in K}} \frac{\|f(x) - T(x)\|}{\|x\|} = 0.$$

In this case we say that T is an asymptotic derivative of f with respect to K *or* a derivative at infinity with respect to K.

If K is a generating cone in E and $T \in \mathcal{L}(E, F)$ is an asymptotic derivative, then T is unique. In this case we denote the linear mapping T by f_K^∞. Now, we suppose that $F = E$ and $K \subset E$ is a generating, closed pointed convex cone. We say that a linear mapping $T : E \to E$ is *positive* if $T(K) \subseteq K$. Similarly, we say that a general mapping $f : E \to E$ is *positive* if $f(K) \subseteq K$.

PROPOSITION 2.2 *If $f : E \to E$ is a positive and asymptotically linear mapping, along the cone K, then f_K^∞ is a positive linear mapping.*

Proof. Suppose that there exists $x_* \in K$ such that $f_K^\infty(x_*) \notin K$. Without restriction we may suppose that $\|x_*\| = 1$. By the formula used in Definition 2.9 we have

$$\lim_{a \to \infty} \frac{f(ax_*)}{a} = f_K^\infty(x_*),$$

and by the closedness of K we have $f_K^\infty(x_*) \in K$ and we have a contradiction. Therefore, $f_K^\infty(K) \subseteq K$; that is, f_K^∞ is positive. $\qquad\square$

LEMMA 2.10 *If $(E, \| \cdot \|)$ is a Banach space ordered by a pointed generating closed convex cone $K \subset E$, then there exists a constant $M > 0$ such that for any element $x \in E$, there exist $u, v \in K$ such that $x = u - v$ and $\|u\| \le M\|x\|$, $\|v\| \le M\|x\|$.*

Proof. A proof of this result is given on p. 102 of the Russian edition of Krasnoselskii [1964a]. $\qquad\square$

We recall that a mapping $f : E \to E$ is *completely continuous with respect to K* if f is continuous and for any bounded set $D \subset K$, we have that $f(D)$ is relatively compact. A mapping can be completely continuous with respect to K but not completely continuous with respect to the space E.

THEOREM 2.11 *Let $(E, \| \cdot \|)$ be a Banach space ordered by a generating, closed pointed convex cone $K \subset E$. Let $f : E \to E$ be a completely continuous* *mapping with respect to K. If f is asymptotically linear along the cone K, then f_K^∞ is a linear completely continuous mapping.*

Proof. Because K is a generating cone, then by Lemma 2.10, there exists a constant $M_0 > 0$ such that every $x \in E$ has a decomposition of the form $x = u - v$, with $u, v \in K$ and such that $\|u\| + \|v\| \leq M\|x\|$. Considering this fact it is easily seen that it suffices to show that f_K^∞ maps $\overline{B}(0, 1) \cap K$ into a compact set. Suppose that this is not true. In this case we can suppose that there exists $\varepsilon > 0$ and a sequence $\{x_n\}_{n \in \mathbb{N}} \subset S(0, 1) \cap K$ such that

$$\|f_K^\infty(x_n - x_m)\| > 3\varepsilon$$

for $n \neq m$. Let $\alpha > 0$ be a real number such that for all $x \in K$ with $\|x\| = \alpha$,

$$\|f(x) - f_K^\infty(x)\| < \varepsilon\|x\|$$

(we used the definition of f_K^∞). Then for $n \neq m$ we have

$$\|f(\alpha x_n) - f(\alpha x_m)\| \geq \alpha\|f_K^\infty(x_n - x_m)\|$$
$$- \|f(\alpha x_n) - f_K^\infty(\alpha x_n)\| - \|f(\alpha x_m) - f_K^\infty(\alpha x_m)\| \geq \alpha\varepsilon,$$

which contradicts the compactness of f on bounded subsets of K. $\qquad\square$

It is well known that the asymptotic derivative of a nonlinear mapping, with respect to a Banach space or with respect to a closed convex cone has many interesting applications to the study of bifurcation problems or to the study of fixed points. About this subject the reader is referred to Amann [1973, 1974a,b, 1976], Cac and Gatica [1979], Krasnoselskii [1964a,b], Krasnoselskii and Zabreiko [1984], Talman [1973] among others. We also note that the notion of asymptotic derivative inspired some ideas developed in Mininni [1977]. Now, we cite the following result. We denote by $\rho(f_K^\infty)$ the spectral radius of f_K^∞.

THEOREM 2.12 (KRASNOSELSKII) *Let $(E, \|\cdot\|)$ be a Banach space ordered by a generating, closed pointed convex cone. Let $f : E \to E$ be a positive completely continuous mapping. If f is asymptotically linear and $\rho(f_K^\infty) < 1$, then f has a fixed point in K.*

Proof. A proof of this result can be found in Amann [1974a] and Krasnoselskii [1964a]. Several authors generalized this result, but in this chapter we give another generalization following another point of view and using the asymptotic scalar derivatives. $\qquad\square$

Let $(H, \langle \cdot, \cdot \rangle)$ be a Hilbert space, $\|\cdot\|$ the norm generated by $\langle \cdot, \cdot \rangle$, and $f : H \to H$. We again use the notion of ψ-additivity (Definition 2.3).

THEOREM 2.13 *Suppose that $f(tx)$ is continuous in t for each fixed x. If f is ψ-additive and ψ satisfies*

1. $\psi(ts) \leq \psi(t)\psi(s)$, *for all $t, s \in \mathbb{R}_+$;*

2. $\psi(t) < t$, *for all $t > 1$;*

then there exists a linear mapping $T : H \to H$ such that

$$|\langle f(x) - T(x), x \rangle| \leq \frac{2\theta\psi(\|x\|)\|x\|}{2 - \psi(2)}, \tag{2.7}$$

for all $x \in H$. S is another linear mapping satisfying (2.7) iff $T - S$ is skew-adjoint.

Proof. By Theorem 2.4 there exists a unique linear mapping T such that

$$\|f(x) - T(x)\| \leq \frac{2\theta\psi(\|x\|)}{2 - \psi(2)}, \tag{2.8}$$

for all $x \in H$. Moreover, we have $T(x) = \lim_{n\to\infty}(f(2^n x)/2^n)$, for all $x \in H$. Hence, by using the Cauchy inequality in (2.8), we obtain (2.7). Suppose that S is another linear mapping satisfying (2.7). Hence,

$$|\langle T(x) - S(x), x \rangle| \leq |\langle T(x) - f(x), x \rangle| + |\langle f(x) - S(x), x \rangle|$$
$$\leq \frac{4\theta\psi(\|x\|)\|x\|}{2 - \psi(2)}.$$

Then,

$$|\langle T(x) - S(x), x \rangle| = \left| \left\langle \frac{1}{n}T(nx) - \frac{1}{n}S(nx), x \right\rangle \right|$$
$$\leq \frac{\psi(n)}{n} \frac{4\theta\psi(\|x\|)\|x\|}{2 - \psi(2)}.$$

Because $\lim_{n\to\infty}(\psi(n)/n) = 0$, we obtain that $\langle T(x) - S(x), x \rangle = 0$. Thus, $T - S$ is skew-adjoint. Conversely, if $T - S$ is skew-adjoint, then $\langle T(x) - S(x), x \rangle = 0$. Hence,

$$|\langle f(x) - S(x), x \rangle| \leq |\langle f(x) - T(x), x \rangle| + |\langle T(x) - S(x), x \rangle|$$
$$= |\langle f(x) - T(x), x \rangle| \leq \frac{2\theta\psi(\|x\|)\|x\|}{2 - \psi(2)}.$$

$\square$

2.4   Asymptotic Differentiability in Locally Convex Spaces

First we recall the following definition of a locally convex space. Let E be a real vector space. We suppose that in E is defined a family of seminorm $\{|\cdot|_\alpha\}_{\alpha\in\mathcal{A}}$ which generates a topology τ such that E endowed with this topology is a locally convex topological vector space; that is, the collection of sets $\{\{x : |x|_\alpha \leq \lambda\} : \alpha \in \mathcal{A}$ and λ is a positive real number$\}$ is a base for a filter of neighbourhoods of zero in E. About the family $\{|\cdot|_\alpha\}_{\alpha\in\mathcal{A}}$ of seminorms we suppose satisfied the following properties.

(i) $(\forall x \in E)(x \neq 0)(\exists \alpha_0 \in \mathcal{A})(|x|_{\alpha_0} \neq 0),$

(ii) $(\forall \alpha_1, \alpha_2 \in \mathcal{A})(\exists \alpha \in \mathcal{A})(|\cdot|_{\alpha_1}, |\cdot|_{\alpha_2} \leq |\cdot|_\alpha).$

We note that the topology defined on E by the family $\{|\cdot|_\alpha\}_{\alpha\in\mathcal{A}}$ of seminorms is a Hausdorff topology. We denote this locally convex space by $(E, \{|\cdot|_\alpha\}_{\alpha\in\mathcal{A}})$. Let $(E, \{|\cdot|_\alpha\}_{\alpha\in\mathcal{A}})$ and $(F, \{|\cdot|_\beta\}_{\beta\in\mathcal{B}})$ be two locally convex spaces and $f : E \rightarrow F$ a linear mapping. We know (see [Marinescu, 1963]) that f is continuous if and only if there exists a function $\psi : \mathcal{B} \rightarrow \mathcal{A}$ such that $|x|_{\psi(\beta)} = 0$ implies $|f(x)|_\beta = 0$ and

$$|f|_{\beta,\psi(\beta)} := \sup_{|x|_{\psi(\beta)}\neq 0} \frac{|f(x)|_\beta}{|x|_{\psi(\beta)}} < \infty,$$

for every $\beta \in \mathcal{B}$.

If $\mathcal{L}(E, F)$ is the vector space of linear continuous mappings from E into F, then $\mathcal{L}(E, F)$ is the pseudo-topological union of spaces $\mathcal{L}_\psi(E, F)$, where

$$\mathcal{L}_\psi(E, F) = \{f : E \rightarrow F \mid f \text{ is linear and } |f|_{\beta,\psi(\beta)} < +\infty, \text{ for every } \beta \in \mathcal{B}\};$$

that is,

$$\mathcal{L}(E, F) = \bigcup_{\psi\in\mathcal{F}(\mathcal{B},\mathcal{A})} \mathcal{L}_\psi(E, F),$$

where

$$\mathcal{F}(\mathcal{B}, \mathcal{A}) = \{\psi : \mathcal{B} \rightarrow \mathcal{A}\}.$$

For this result the reader is referred to Marinescu [1963]. Let $K \subset E$ be a closed pointed convex cone. We suppose that K is *total* in E; that is, $\overline{K - K} = E$.

DEFINITION 2.14 *We say that a mapping $f : K \rightarrow F$ is asymptotically linear along the cone K if there exist a function $\psi : \mathcal{B} \rightarrow \mathcal{A}$ and a linear continuous mapping $f_\infty \in \mathcal{L}_\psi(E, F)$ such that*

$$\lim_{\substack{x\in K \\ |x|_{\psi(\beta)}}} \frac{|f(x) - f_\infty(x)|_\beta}{|x|_{\psi(\beta)}} = 0$$

for any $\beta \in \mathcal{B}$.

Because the cone K is total in E, the mapping f_∞ is unique. We say that f_∞ is the *asymptotic derivative* of f along the cone K. Let $(E, \{|\cdot|_\alpha\}_{\alpha \in A})$ be an arbitrary locally convex space. For every $\alpha \in A$ and every subset $\Omega \subset E$ we define the measures of noncompactness:

$$\gamma_\alpha(\Omega) = \inf\{d > 0 \ : \ \Omega \text{ can be covered by a finite number of sets of}$$
$$\|\cdot\|_\alpha\text{-diameter} \leq d\},$$

$$\chi_\alpha(\Omega) = \inf\{r > 0 \ : \ \Omega \text{ can be covered by a finite number of } \|\cdot\|_\alpha\text{-balls of}$$
$$\|\cdot\|_\alpha\text{-radius} \leq d\}.$$

We consider the set of functions

$$\mathcal{C} = \{f : A \to [0, +\infty]\}$$

ordered by the ordering

$$f, g \in \mathcal{C}, \ \ f \leq g \text{ if and only if } f(\alpha) \leq g(\alpha) \text{ for any } \alpha \in A.$$

We consider also the following functions: $\gamma : 2^E \to \mathcal{C}$ defined by

$$\Omega \mapsto (\gamma(\Omega))(\alpha) = \gamma_\alpha(\Omega),$$

for any $\alpha \in A$. $\chi : 2^E \to \mathcal{C}$ defined by

$$\Omega \mapsto (\chi(\Omega))(\alpha) = \chi_\alpha(\Omega),$$

for any $\alpha \in A$. Because the functions γ and χ have similar properties we denote by Φ or the function γ or the function χ. In nonlinear analysis it is known that the function Φ has the following properties.

(1) $\Phi(A \cup B) \leq \max\{\Phi(A), \Phi(B)\}$, for any $A, B \in 2^E$.

(2) $\Phi(A) = 0$ if and only if A is a totally bounded set. (We recall that a set A is totally bounded if for each 0-neighbourhood U there exists a finite subset $A_0 \subset A$ such that $A \subset A_0 + U$. Because a locally convex space is Hausdorff, a subset A is totally bounded if and only if it is precompact.)

(3) $\Phi(\overline{co}(A)) = \Phi(A)$, where $\overline{co}(A)$ is the closed convex hull of A.

(4) $A \subseteq B$ implies $\Phi(A) \leq \Phi(B)$.

(5) For any $\lambda \in \mathbb{R}_+$ and any $\alpha \in A$ we have $\Phi(\lambda A) = \lambda \Phi(A)$.

(6) If B_α is the open ball of $|\cdot|_\alpha$-radius $= 1$, then $\Phi(B_\alpha) \leq 2$.

(7) For any $\alpha \in A$ we have $\Phi(A + B) \leq \Phi(A) + \Phi(B)$.

For the proof of properties (1)–(7) the author is referred to references [1], [4], [12–15] and [17] cited in Isac [1982]. Let E and F be locally convex spaces such that the family of seminorms for each space is denoted by the same set $\mathcal{A}$; that is, $(E, \{\|\cdot\|_\alpha\}_{\alpha \in \mathcal{A}})$ and $(F, \{\|\cdot\|_\beta\}_{\beta \in \mathcal{A}})$. Therefore, for both spaces we have the same set $\mathcal{C}$. We note that we have this situation in particular when $E = F$ or when E and F are Frechét spaces. We denote by Φ_E (resp., Φ_F) the function defined above considering the function γ (resp., χ). Let D be a subset of E (supposed to be a nonempty set). We can have $D = E$.

DEFINITION 2.15 *We say that a mapping* $f : D \to F$ *is an* (α^*, Φ)-*contraction if*

$$\Phi_F(f(Q)) \leq \alpha^* \Phi_E(Q),$$

for any nonempty bounded set $Q \subset D$, *where* α^* *is a function from* A *into* $\mathbb{R}_+$.

REMARK 2.5 *Because* $\Phi_E(Q) : \mathcal{A} \to [0, +\infty]$ *the inequality used in Definition 2.15 means*

$$\Phi_F(f(Q))(\alpha) \leq \alpha^*(\alpha)\Phi_E(Q)(\alpha).$$

The following result is given with respect to a total closed convex cone, but we have a similar proof when the cone is the space itself. This result is due to Isac.

THEOREM 2.16 *Let* E *and* F *be locally convex spaces such that the family of seminorms for each space is indexed by the same set* $\mathcal{A}$; *that is,* $(E, \{\|\cdot\|_\alpha\}_{\alpha \in \mathcal{A}})$ *and* $(F, \{\|\cdot\|_\beta\}_{\beta \in \mathcal{A}})$. *Let* $K \subset E$ *be a total closed convex cone and* $f : K \to F$ *a* (α^*, Φ)-*contraction mapping. If* f *is asymptotically linear along the cone* K, *then* $f_\infty|_K$ *is an* (α^*, Φ)-*contraction.*

Proof. We denote $u = f_\infty$. Let $\alpha \in A$ be an arbitrary element and $A \subset K$ a bounded subset such that there exists $\rho > 0$ with the property that $|x|_{\psi(\alpha)} \geq \rho$, for any $x \in A$ (the function ψ is given by Definition 2.15). Let σ be a positive real number such that

$$\sigma > \sup\{|x|_{\psi(\alpha)} : x \in A\},$$

and let $\varepsilon > 0$ be arbitrary. We denote $r = f - u$. Because f is asymptotically linear along the cone K, there exists $\delta > 0$ such that for any $x \in K$ with the property $|x|_{\psi(\alpha)} \geq \delta$ we have

$$|r(x)|_\alpha \leq \frac{\varepsilon}{2\sigma}.$$

If we denote $B_\alpha^F = \{x \in F : |x|_\alpha < 1\}$, then for any $\lambda \geq (\delta/\rho)$ we have

$$r(\lambda A) \subset \frac{\lambda \varepsilon}{B_\alpha^F},$$

because

$$|\lambda x|_{\psi(\alpha)} = \lambda |x|_{\psi(\alpha)} \geq \frac{\delta}{\rho} \|x\|_{\psi(\alpha)} \geq \delta.$$

Therefore, we have

$$u(\lambda A) \subset f(\lambda A) - r(\lambda A) \subset f(\lambda A) + \frac{\lambda \varepsilon}{2} B_\alpha^F,$$

which implies (denoting by γ^E (resp., γ^F) the measure of noncompactness on E (resp., on F))

$$\lambda \gamma_\alpha^F(u(A)) = \gamma_\alpha^F(u(\lambda A)) \leq \gamma_\alpha^F(f(\lambda A)) + \frac{\lambda \varepsilon}{2} \gamma_\alpha^F(B_\alpha^F)$$

$$\leq \alpha^*(\alpha) \gamma_\alpha^E(\lambda A) + \lambda \varepsilon = \lambda(\alpha^*(\alpha) \gamma_\alpha^E(A) + \varepsilon).$$

Because $\varepsilon > 0$ is arbitrary we have

$$\gamma_\alpha^F(u(A)) \leq \alpha^*(\alpha) \gamma_\alpha^F(A). \tag{2.9}$$

Now, we suppose that $A \subset K$ is an arbitrary nonempty bounded set and $\varepsilon > 0$ is an arbitrary real number. Because $u = f_\infty \in \mathcal{L}_\psi(E, F)$, where $\psi : \mathcal{A} \to \mathcal{A}$ is used for the continuity of U and for the asymptotic linearity of f we have,

$$|u|_{\alpha,\psi(\alpha)} = \sup_{|x|_{\psi(\alpha)} \neq 0} \frac{|u(x)|_\alpha}{|x|_{\psi(\alpha)}} < \infty,$$

and if $|x|_{\psi(\alpha)} = 0$, then $|u(x)|_\alpha = 0$. For the properties of $|u|_{\alpha,\psi(\alpha)}$ see Marinescu [1963]. First, we suppose that $|u|_{\alpha,\psi(\alpha)} \neq 0$ and we take

$$\rho = \frac{\varepsilon}{2|u|_{\alpha,\psi(\alpha)}}.$$

We define $A_1 = A \cap \rho B_{\psi(\alpha)}^E$ and $A_2 = A \backslash A_1$. In this case we have

$$u(A_1) \subset \frac{\varepsilon}{2} B_\alpha^F. \tag{2.10}$$

If $|u|_{\alpha,\psi(\alpha)} = 0$, we take $A_1 = A \cap B_{\psi(\alpha)}^E$ and $A_2 = A \backslash A_1$ and considering the definition of $|u|_{\alpha,\psi(\alpha)}$ and the continuity of u we obtain again the formula (2.10). In both situations we have

$$\gamma_\alpha^F(u(A_1)) \leq \varepsilon.$$

From the first part of the proof we deduce

$$\gamma_\alpha^F(u(A_2)) \leq \alpha^*(\alpha) \gamma_\alpha^E(A_2) \leq \alpha^*(\alpha) \gamma_\alpha^E(A).$$

Finally, we obtain

$$\gamma_\alpha^F(u(A)) = \gamma_\alpha^F(u(A_1) \cup u(A_2)) \leq \max\{\gamma_\alpha^F(u(A_1)), \gamma_\alpha^F(u(A_2))\}$$
$$\leq \max\{\varepsilon, \alpha^*(\alpha)\gamma_\alpha^E(A)\}.$$

Because $\varepsilon > 0$ is arbitrary, we obtain formula (2.9) for an arbitrary bounded set $A \subset K$. Now, if we pass to the function Φ we have

$$\Phi_F(u|_K(A)) \leq \alpha^* \Phi_E(A),$$

because the same proof is valid if we replace for any $\alpha \in \mathcal{A}$ the measure of noncompactness γ_α by χ_α. The proof of the theorem is complete. $\qquad\square$

We recall that A. Granas defined the notion of quasi-bounded mapping Granas [1962]. Let $(E, \|\cdot\|)$ be a Banach space and $f : E \to E$ a mapping. We say that f is a *quasi-bounded mapping* if

$$\limsup_{\|x\|\to\infty} \frac{\|f(x)\|}{\|x\|} = \inf_{\rho>0} \sup_{\|x\|\geq\rho} \frac{\|f(x)\|}{\|x\|} < \infty.$$

If f is quasi-bounded, then the real number

$$|f|_{qb} = \limsup_{\|x\|\to\infty} \frac{\|f(x)\|}{\|x\|}$$

is called the *quasi-norm* of f. Any bounded linear mapping $L : E \to E$ is quasi-bounded and $|L|_{qb} = \|L\|$. If f is a nonlinear mapping such that $\exists M > 0$ with the property

$$\|f(x)\| \leq M\|x\|,$$

for any $x \in E$, then f is quasi-bounded. The notion of quasi-bounded mapping has interesting applications in fixed point theory. Now, we generalize this notion to locally convex spaces. Let $(E, \{|\cdot\|_\alpha\}_{\alpha\in\mathcal{A}})$ and $(F, \{|\cdot\|_\beta\}_{\beta\in\mathcal{B}})$ be two totally convex spaces and $f : E \to F$ a mapping.

DEFINITION 2.17 *We say that f is quasi-bounded if there exists a function $\psi : \mathcal{B} \to \mathcal{A}$ such that the numbers*

$$\|f\|_{\beta,\psi(\beta)} = \inf_{0<\rho<\infty} \left\{ \sup_{|x|_{\psi(\beta)}\geq\rho} \frac{|f(x)|_\beta}{|x|_{\psi(\beta)}} \right\}$$

are finite for any $\beta \in \mathcal{B}$.

The following results are consequences of Definition 2.17.

PROPOSITION 2.3 *If the mapping $f : E \to F$ is quasi-bounded, then there exists a function $\psi : \mathcal{B} \to \mathcal{A}$ such that for any $\varepsilon > 0$ there exists $\rho > 0$ with the property that for any $x \in E$ with $|x|_{\psi(\beta)} > \rho$, we have*

$$|f(x)|_\beta \leq (\|f\|_{\beta,\psi(\beta)} + \varepsilon)|x|_{\psi(\beta)}.$$

PROPOSITION 2.4 *If there exists a function* $\psi : \mathcal{B} \to \mathcal{A}$ *and for each* $\beta \in \mathcal{B}$ *there exist two constants* $k_{1,\psi(\beta)} \geq 0$ *and* $k_{2,\psi(\beta)} \geq 0$ *such that*

$$|f(x)|_\beta \leq k_{1,\psi(\beta)}|x|_{\psi(\beta)} + k_{2,\psi(\beta)},$$

for any $x \in E$, *then* f *is a quasi-bounded mapping and* $\|f\|_{\beta,\psi(\beta)} \leq k_{1,\psi(\beta)}$.

We recall that if $T : E \to F$ is a linear mapping, then T is continuous if for any $\beta \in \mathcal{B}$ there exists $\alpha \in \mathcal{A}$ and a constant $M_{\beta,\alpha} \geq 0$ such that

$$|T(x)|_\beta \leq M_{\beta,\alpha}|x|_\alpha,$$

for any $x \in E$. For any $T \in \mathcal{L}(E, F)$, $\alpha \in \mathcal{A}$, and $\beta \in \mathcal{B}$ we define

$$|T|_{\beta,\alpha} = \sup_{|x|_\alpha \leq 1} |T(x)|_\beta.$$

We observe that $|T|_{\beta,\alpha}$ can be zero for $T \neq 0$ and it can be $+\infty$. We have that $T \in \mathcal{L}(E, F)$ is continuous if and only if there exists a function $\psi : \mathcal{B} \to \mathcal{A}$ such that

$$T \in \mathcal{L}_\psi(E, F) = \{f : E \to F : f \text{ is linear and } |f|_{\beta,\psi(\beta)} < +\infty \text{ for any}$$
$$\beta \in \mathcal{B}\}.$$

PROPOSITION 2.5 *If* $f : E \to F$ *is asymptotically linear along the cone* $K = E$, *then* f *is quasi-bounded and*

$$\|f\|_{\beta,\psi(\beta)} = |f_\infty|_{\beta,\psi(\beta)},$$

where ψ *is the function used in the definition of asymptotic linearity and in the continuity of* f_∞.

In 1973, Louis A. Talman, presented in his PhD thesis (Graduate School of the University of Kansas) another approach of asymptotical differentiability along a closed convex cone in a locally convex space [Talman, 1973]. Now we present his approach and some of his results. Let $(E, \{|\cdot|_\alpha\}_{\alpha \in \mathcal{A}})$ be an arbitrary (Hausdorff) locally convex space. Consider again the power set 2^E and the set $\mathcal{C}(\mathcal{A}) = \mathcal{C} = \{f : \mathcal{A} \to [0, +\infty]\}$ ordered by $f \leq g$ if and only if $f(\alpha) \leq g(\alpha)$, for any $\alpha \in \mathcal{A}$.

DEFINITION 2.18 *We say that a function* $\Psi : 2^E \to \mathcal{C}(\mathcal{A})$ *is a measure of noncompactness on* E *if for every* $A, B \in 2^E$, *for every* $\lambda \in \mathbb{R}$, *and for every* $\lambda \in \mathcal{A}$ *the following properties are satisfied.*

(1) $\Psi(A)(\alpha) < +\infty$ *if* A *is bounded.*

(2) $\Psi(A) \equiv 0$ *if and only if* A *is precompact.*

(3) $\Psi(A \cup B) \leq \max(\Psi(A), \Psi(B))$.

(4) $\Psi(\lambda A) = |\lambda|\psi(A)$.

(5) $\Psi(A + B) \leq \Psi(A) + \Psi(B)$.

(6) $\Psi(A) = \Psi(\mathrm{cl}_E A) = \Psi(\mathrm{conv}\, A)$, *where* $\mathrm{cl}_E A$ *is the closure of* A *with respect to* E *and* $\mathrm{conv}\, A$ *is the convex hull of* A.

(7) *There is a convex balanced neighbourhood of zero,* U_α, *in* E *such that* $\Psi(U_\alpha) = 1$ *and such that if* $\Psi(A)(\alpha) \leq \rho < \infty$, *then for every* $\delta > 0$ *there is a finite set* $\{x_1, x_2, \ldots, x_n\} \subseteq E$ *with the property that*

$$A \subseteq \bigcup_{k=1}^{n} [x_k + (\rho + \delta)U_\alpha].$$

From property (3) we deduce that a measure of noncompactness is monotone; that is, if $A \subseteq B$, then $\Psi(A) \leq \Psi(B)$. Also, a consequence of properties (2) and (5) is the fact that a measure of noncompactness is translation invariant. Indeed, we have

$$\Psi(A) = \Psi((x_0 + A) - x_0) \leq \Psi(x_0 + A) + \Psi(-x_0) = \Psi(x_0 + A)$$
$$\leq \Psi(x_0) + \Psi(A) = \Psi(A),$$

so that $\Psi(x_0 + A) = \Psi(A)$. For more information and results about measures of noncompactness the reader is referred to Sadovskii [1968] and Banas and Goebel [1980]. The notion of noncompactness defined above includes the two most commonly used measures of noncompactness, namely, γ (the Kuratowski measure of noncompactness) and χ (the Hausdorff measure of noncompactness) defined in this section of this chapter.

Let $\Psi : 2^E \to \mathcal{C}(\mathcal{A})$ be a measure of noncompactness, $M \subset E$ a nonempty set, and $f : M \to E$ a mapping.

DEFINITION 2.19 *We say that* f *is a* k-Ψ-*contraction if there is a function* $k : \mathcal{A} \to [0, 1[$ *such that*

$$\Psi(f(B))(\alpha) \leq k(\alpha)\Psi(\beta)(\alpha),$$

for every bounded set $B \subseteq M$ *and for every* $\alpha \in \mathcal{A}$.

It is known that there exists an extensive literature concerning k-Ψ-contractions in Banach spaces. For locally convex spaces we cite Talman [1973].

Let $K \subset E$ be a closed pointed convex cone. We recall that K is *total* in E if $E = \overline{K - K}$ and K is *generating* if $E = K - K$.

DEFINITION 2.20 *We say that K is* sharp *(or* locally bounded*) if there is a neighbourhood U of zero in E such that $K \subset \partial_E U$ is bounded and nonempty. (We denote by $\partial_E U$ the boundary of U with respect to E.)*

In any Banach space any closed pointed convex cone is sharp. In a locally convex space we can show that if U is a neighbourhood of zero for which $K \cap \partial_E U$ is bounded, then $K \cap U$ is also bounded. Talman has given an example of a sharp cone in a non-normed vector space, which is also a generating cone. Now we present his example.

Let $(E, \| \cdot \|)$ be a Banach space, E^* its topological dual, and suppose that the $\sigma(E, E^*)$ topology on E (the weak topology) is not a norm topology. We select a nonzero element $x_* \in E^*$ and choose $x_0 \in E$ with the property $\langle x_*, x_0 \rangle = 1$. We define

$$K = \{x \in E \ : \ \|x - \langle x_*, x \rangle x_0\| \le \langle x_*, x \rangle\}.$$

(We note that $\| \cdot \|$ is the norm on E and $\langle \cdot, \cdot \rangle$ is the bilinear form which defines the duality between E^* and E.) The set K is a convex cone. Indeed, if $x, y \in K$, then

$$\begin{aligned} \|(x + y) - \langle x_*, x + y \rangle x_0\| &\le \|x - \langle x_*, x \rangle x_0\| + \|y - \langle x_*, y \rangle x_0\| \\ &\le \langle x_*, x \rangle + \langle x_*, y \rangle = \langle x_*, x + y \rangle, \end{aligned}$$

which implies that $x + y \in K$.

Obviously, if $\lambda \ge 0$ and $x \in K$, then $\lambda x \in K$. We also remark that $K \ne \{0\}$ because $x_0 \in K$. Because K is convex and closed in the norm topology, K is closed for $\sigma(E, E^*)$. Now, we show that K is pointed; that is, $K \cap (-K) = \{0\}$. Indeed, if $x \in K$ and $-x \in K$, then we have

$$0 \le \| \pm x - \langle x_*, \pm x \rangle x_0\| \le \langle x_*, \pm x \rangle,$$

which implies that $\langle x_*, x \rangle = 0$. Thus,

$$0 = \|x - \langle x_*, x \rangle x_0\| = \|x\|,$$

and $\|x\| = 0$. Hence, K is a pointed convex cone. Let

$$U = \{x \in E \ : \ |\langle x_*, x \rangle| \le 1.$$

Then, U is a neighbourhood of zero for $\sigma(E, E^*)$ and we can show that

$$\partial_E^\sigma U = \{x \in E \ : \ |\langle x_*, x \rangle| = 1\}$$

(here of course $\partial_E^\sigma U$ means the boundary mapping for the weak topology $\sigma(E, E^*)$). Hence,

$$K \cap \partial_E^\sigma U = \{x \in E \ : \ |\langle x_*, x \rangle| = 1 \text{ and } \|x - x^*\| \le 1\},$$

which is clearly bounded for the norm topology and therefore is bounded for $\sigma(E, E^*)$. Finally, we show that K is a generating cone in E. Indeed, let $x \in E$ be an arbitrary element. We must show that $x \in K - K$, and we must assume that $x \notin K$. Put $\alpha = \langle x_*, x \rangle$, and let $\beta = \|x - \alpha x_0\|$. Because $x \notin K$, we have $\beta - \alpha > 0$. Let $y = x + (\beta - \alpha)x_0$. Then, we have

$$\|y - \langle x_*, y \rangle x_0\| = \|x + (\beta - \alpha)x_0 - \langle x_*, x + (\beta - \alpha)x_0 \rangle\|$$
$$= \|x + (\beta - \alpha)x_0 - \langle x_*, x \rangle x_0 - \langle x_*, (\beta - \alpha)x_0 \rangle x_0\|$$
$$= \|x + (\beta - \alpha)x_0 - \alpha x_0 - (\beta - \alpha)x_0\| = \|x - \alpha x_0\| = \beta = \alpha + (\beta - \alpha)$$
$$= \langle x_*, x \rangle + (\beta - \alpha)\langle x_*, x_0 \rangle = \langle x_*, x + (\beta - \alpha)x_0 \rangle = \langle x_*, y \rangle.$$

Therefore, $y \in K$. Because $\beta - \alpha > 0$ and $x_0 \in K$, we know that $(\beta - \alpha)x_0 \in K$. Then, $x = y - (\beta - \alpha)x \in K - K$ and we have that $E = K - K$.

Let $(E, \{p_\alpha\}_{\alpha \in A})$ be a Hausdorff locally convex space and $K \subset E$ a closed pointed convex cone. We denote by $\mathcal{L}(E, E)$ the vector space of linear continuous mappings from E into E.

DEFINITION 2.21 *We say that a mapping $f : K \rightarrow E$ is Hyers–Lang asymptotically linear (HLAL) along K if there is a continuous linear mapping $\mathcal{D}_\infty f : E \rightarrow E$ (i.e., $\mathcal{D}_\infty f \in \mathcal{L}(E, E)$) such that for any $\alpha, \beta \in A$ there exist $\gamma \in A$ and constants $c_\alpha, c_\beta > 0$ such that the following properties are satisfied.*

1. *$p_\alpha(x) \leq c_\alpha p_\gamma(x)$, for any $x \in K$.*

2. *For any $\varepsilon > 0$ there exists $M > 0$ with the property that $x \in K$ and $p_\gamma(x) \geq M$ imply*

$$p_\beta[f(x) - \mathcal{D}_\infty f(x)] \leq \varepsilon c_\beta p_\gamma(x).$$

REMARK 2.6 *In Definition 2.21 we can suppress the constants c_α and c_β. We can do this if we denote again by p_γ the seminorm cp_γ, where $c = \max\{c_\alpha, c_\beta\}$.*

Now, we recall a well known notion in the theory of locally convex spaces. Let $A \subset E$ be a nonempty subset. We say that A is *balanced* (*circled*) if $\lambda A \subseteq A$, whenever $|\lambda| \leq 1$ and we say that A is *radial* (*absorbing*), if for each $x \in E$ there is an $\varepsilon > 0$ such that $tx \in A$ for $t \in [0, \varepsilon]$. If $B \subset E$ is a *radial*, *balanced*, and *convex* set, then the nonnegative real function

$$x \mapsto p_B(x) = \inf\{\lambda > 0 : x \in \lambda B\}$$

is called the *Minkowski functional* associated with B. In this case we can show that p_B is a seminorm. We recall that a subset $A \subset E$ is bounded if for each 0-neighbourhood $U \in E$, there exists $\lambda \in R$ such that $A \subset \lambda U$. If $B \subset E$ is radial, balanced, convex, and bounded, then in this case p_B is a norm on E.

We denote by E_B the normed space obtained by equipping the linear subspace of E that is spanned by B with the norm arising from the Minkowski functional p_B. If q is any continuous seminorm on E, we let E_q denote the quotient space $E/\ker(q)$ equipped with the canonical norm induced by q. (We recall that $\ker(q) = \{x \in E : q(x) = 0\}$.) We denote by $\pi_q : E \to E_q$ the quotient mapping. If $u : E \to E$ is a continuous linear mapping, then $\pi_q \circ u|_{E_B} : E_B \to E_q$ is continuous.

If $q = p_\alpha$, we denote the space E_{p_α} by E_α. Let $\mathcal{B}$ be a basis consisting of closed, balanced, absorbing, convex sets for the bornology on E.

DEFINITION 2.22 *We say that a mapping* $f : K \to E$ *is* $\mathcal{B}$*-asymptotically linear (*$\mathcal{B}$*AL) along* K *if there is a continuous linear mapping* $\mathbf{D}_\infty f : E \to E$ *such that for every* $\alpha \in \mathcal{A}$ *and for every* $B \in \mathcal{B}$, $\pi_\alpha \circ \mathbf{D}_\infty f|_{E_B} : E_B \to E_\alpha$ *is the asymptotic derivative (in the sense of normed spaces along* $K \cap E_B$ *of the mapping* $\pi_\alpha \circ f|_{E_B} : E_B \to E_\alpha$*).*

REMARK 2.7 *The notion of* $\mathcal{B}$*-asymptotic linearity is highly sensitive to the selection of* $\mathcal{B}$*.*

PROPOSITION 2.6 *Let* $(E, \{p_\alpha\}_{\alpha \in \mathcal{A}})$ *be a Hausdorff locally convex space and* $K \subset E$ *a total closed pointed convex cone. If* $f : K \subset E$ *is HLAL along* K *with HL-asymptotic derivative* $\mathcal{D}_\infty f$ *and for some basis* $\mathcal{B}$ *for the bornology of* E, f *is* $\mathcal{B}$*AL along* K *with* $\mathcal{B}$*-asymptotic derivative* $\mathbf{D}_\infty f$*, then* $\mathbf{D}_\infty f = \mathcal{D}_\infty f$*.*

Proof. Let $x_0 \in K$ be an arbitrary element such that $x_0 \neq 0$. Let $B \in \mathcal{B}$ be such that $x_0 \in B$ and let $\alpha \in \mathcal{A}$ be arbitrary. Then, for every $\lambda > 0$ we have

$$\frac{p_\alpha(\mathbf{D}_\infty f(\lambda x_0) - \mathcal{D}_\infty f(\lambda x_0))}{p_B(\lambda x_0)}$$
$$\leq \frac{p_\alpha(\mathbf{D}_\infty f(\lambda x_0) - f(\lambda x_0)) + p_\alpha(f(\lambda x_0) - \mathcal{D}_\infty f(\lambda x_0))}{p_B(\lambda x_0)}. \tag{2.11}$$

Let $\varepsilon > 0$ be given. Because f is HLAL, there is a $\beta \in \mathcal{A}$ such that $p_\beta(x_0) \neq 0$, and such that $\lambda > 0$ sufficiently large implies that

$$\frac{p_\alpha(f(\lambda x_0) - \mathcal{D}_\infty f(\lambda x_0))}{p_\beta(\lambda x_0)} < \frac{\varepsilon p_B(x_0)}{2 p_\beta(x_0)}.$$

(We note that $p_B(x_0) \neq 0$, because otherwise $\lambda x_0 \in B$ for all $\lambda > 0$, which is impossible because B is bounded.) It now follows that, for large $\lambda > 0$, we have

$$\frac{p_\alpha(f(\lambda x_0) - \mathcal{D}_\infty f(\lambda x_0))}{p_B(\lambda x_0)} = \frac{p_B(x_0) p_\alpha(f(\lambda x_0) - \mathcal{D}_\infty f(\lambda x_0))}{p_B(\lambda x_0) p_B(x_0)} < \frac{\varepsilon}{2}.$$

On the other hand f is $\mathcal{B}$AL, so that when $\lambda > 0$ is large, we have

$$\frac{p_\alpha(\mathbf{D}_\infty f(\lambda x_0) - f(\lambda x_0))}{p_B(\lambda x_0)} < \frac{\varepsilon}{2}.$$

Combining the last two inequalities with 2.4, we obtain , for $\lambda > 0$ sufficiently large,

$$\frac{p_\alpha(\mathbf{D}_\infty f(\lambda x_0) - \mathcal{D}_\infty f(\lambda x_0))}{p_B(\lambda x_0)} < \varepsilon.$$

Inasmuch as $\mathbf{D}_\infty f$ and $\mathcal{D}_\infty f$ are both linear this is equivalent to

$$p_\alpha(\mathbf{D}_\infty f(x_0) - \mathcal{D}_\infty f(x_0)) < \varepsilon p_B(x_0).$$

It follows that

$$\mathbf{D}_\infty f(x_0) - \mathcal{D}_\infty f(x_0) \in \ker p_\alpha,$$

and because $\alpha \in \mathcal{A}$ was arbitrary and E is Hausdorff, we must have

$$\mathbf{D}_\infty f(x_0) = \mathcal{D}_\infty f(x_0).$$

Taking into account the totality of K, the proof is complete. $\qquad\square$

REMARK 2.8 *In Talman [1973] are given several examples to show that in general neither of the implications $\mathcal{B}$AL $\implies$ HLAL and HLAL $\implies$ $\mathcal{B}$AL is true, but if some special conditions are satisfied we have an interesting relation between $\mathcal{B}$AL and HLAL along K.*

A basis $\mathcal{B}$ for the bornology of a locally convex space E is said to be *strict* if each $B \in \mathcal{B}$ has the property that for every bounded set $D \subset E$, B absorbs $D \cap E_B$. It seems that there is no topological condition which guarantees that such a basis exists. There exist spaces in which there is no such basis [Talman, 1973].

THEOREM 2.23 *Let $(E, \{p_\alpha\}_{\alpha \in \mathcal{A}})$ be a locally convex space whose bornology admits a strict basis $\mathcal{B}$ and let K be a sharp cone in E. Then, $f : K \to E$ is $\mathcal{B}$AL along K if and only if f is HLAL along K.*

Proof. Assume that f is $\mathcal{B}$AL along K. Let $\alpha, \beta \in \mathcal{A}$ be given. Find $\theta \in \mathcal{A}$ so that $B_\theta \cap K$ is bounded, where B_θ is the open unit p_θ-ball. Choose $\gamma \in \mathcal{A}$ such that $\max\{p_\alpha, p_\theta\} \le p_\gamma$. Then $B_\gamma \cap K \subseteq B$ for some $B \in \mathcal{B}$ and $p_\beta \le p_\gamma$ on K. But B_γ is a neighbourhood of zero, so B_γ absorbs B, and this means that there is a $k > 0$ such that $p_\gamma \le k p_\beta$ on E_B, which contains K.

If $\varepsilon > 0$ is given, we find M so that whenever $x \in K$ and $p_B(x) \ge M$, we have $p_\beta(f(x) - \mathbf{D}_\infty f(x)) \le \varepsilon p_B(x)$. If $x \in K$ and $p_\gamma(x) \ge kM$, then $p_B(x) \ge M$, so that

$$p_\beta(f(x) - \mathbf{D}_\infty f(x)) \le \varepsilon p_B(x) \le \varepsilon p_\gamma(x).$$

It follows that f is HLAL along K.

Conversely, assume that f is HLAL along K, and let $\alpha \in \mathcal{A}$, $\beta \in \mathcal{B}$ be given. We take β so that $p_\beta \cap K$ is bounded. Find γ so that $p_\beta \leq p_\gamma$, and for every $\varepsilon > 0$ there is an M such that $x \in K$ and $p_\gamma(x) \geq M$ imply that $p_\alpha(f(x) - \mathcal{D}_\infty f(x)) < \varepsilon p_\gamma(x)$. Because $\mathcal{B}$ is strict, there is a $k_1 > 0$ such that $p_B \leq k_1 p_\gamma$ on K (because K is sharp and $B_\gamma \cap K$ is bounded).

B_γ is a neighbourhood of zero, and thus there is a $k_2 > 0$ such that $p_\gamma \leq k_2 p_B$ on E_B. Let $\varepsilon > 0$ be given. Find M so that $p_\gamma(x) \geq M$ implies that

$$p_\alpha(f(x) - \mathcal{D}_\infty f(x)) \leq \frac{\varepsilon}{k_2} p_\gamma(x).$$

Then $p_B(x) \geq k_1 M$ implies that $p_\gamma(x) \geq M$, which in turns implies that

$$p_\alpha(f(x) - \mathcal{D}_\infty f(x)) \leq \frac{\varepsilon}{k_2} p_\gamma(x) \leq \varepsilon p_B(x).$$

Hence, f is $\mathcal{B}$AL along K. $\qquad\qquad\square$

PROPOSITION 2.7 *If $f : K \to K$ is HLAL along K, then $\mathcal{D}_\infty f(K) \subseteq K$.*

Proof. We suppose that there is an $h \in K$ such that $\mathcal{D}_\infty f(h) \notin K$. We select a positive $x_* \in E^*$ such that

$$\langle x_*, \mathcal{D}_\infty f(h) \rangle < 0.$$

We denote

$$\mu = \langle x_*, \mathcal{D}_\infty f(h) \rangle$$

and we define

$$\phi(t) = \langle x_*, f(th) \rangle.$$

If $t \geq 0$, then $th \in K$, so that $f(th) \in K$ and $\phi(t) \geq 0$. But for $t > 0$,

$$\phi(t) = t \left[\left\langle x_*, \frac{f(th) - \mathcal{D}_\infty f(th)}{t} \right\rangle + \mu \right],$$

and the function

$$x \mapsto |\langle x_*, x \rangle|$$

is a continuous semi-norm on E. Hence, there is an $\alpha \in \mathcal{A}$ such that

$$|\langle x_*, x \rangle| \leq p_\alpha(x)$$

(modulo a multiplicative constant, which we ignore). Because E is Hausdorff, there is a $\beta \in \mathcal{A}$ with $K \setminus \ker p_\beta \neq \emptyset$. Select $\gamma \in \mathcal{A}$ so that $p_\beta \leq p_\gamma$ and so that for every $\varepsilon > 0$ there is an $M > 0$ such that $t \geq 0$ and $p_\gamma(th) \geq M$ imply that

$$p_\alpha(f(th) - \mathcal{D}_\infty f(th)) \leq \varepsilon p_\gamma(th).$$

If we now choose ε so that

$$0 < \varepsilon < -\frac{\mu}{2p_\gamma(h)}$$

and find M accordingly we see that when

$$t \geq \frac{M}{p_\gamma(h)}$$

we must have

$$\left| \left\langle x_*, \frac{f(th) - \mathcal{D}_\infty f(th)}{t} \right\rangle \right| \leq p_\alpha \left(\frac{f(th) - \mathcal{D}_\infty f(th)}{t} \right)$$

$$\leq \frac{1}{t} \varepsilon p_\gamma(th) = \varepsilon p_\gamma(h) < -\frac{\mu}{2}.$$

But then, for such t we would have

$$\phi(t) = t \left[\left\langle x_*, \frac{f(th) - \mathcal{D}_\infty f(th)}{t} \right\rangle + \mu \right] < \frac{\mu t}{2} < 0,$$

which is not possible because $\phi(t) \geq 0$. This contradiction completes the proof.
□

PROPOSITION 2.8 *If $f : K \to K$ is $\mathcal{B}AL$ along K for some basis $\mathcal{B}$ for the bornology of E, then $\mathcal{D}_\infty f(K) \subseteq K$.*

Proof. With $\mathbf{D}_\infty f$ in place of $\mathcal{D}_\infty f$, we proceed as in the proof of Proposition 2.7 up through the definition of the function $\phi : \mathbb{R} \to \mathbb{R}$. Select a $B \in \mathcal{B}$ so that $\{h\} \subseteq H$. Because $x \mapsto |\langle x_*, x \rangle|$ is a continuous semi-norm on E, there is an $M > 0$ such that $x \in E_B \cap K$ and $p_B(x) \geq M$ imply that

$$|\langle x_*, f(x) - \mathbf{D}_\infty f(x) \rangle| < -\frac{\mu}{2} \frac{p_B(x)}{p_B(h)}.$$

But then when

$$t > \frac{M}{p_B(h)},$$

we have $p_B(th) > M$, so that, for such t,

$$\frac{1}{t} |\langle x_*, f(th) - \mathbf{D}_\infty f(th) \rangle| < -\frac{1}{t} \frac{\mu}{2} \frac{p_B(th)}{p_B(h)} = -\frac{\mu}{2}$$

and again $\phi(t) < 0$. The proof is complete.
□

THEOREM 2.24 *Let* $(E, \{p_\alpha\}_{\alpha \in \mathcal{A}})$ *be a Hausdorff locally convex space,*

$$\Psi : 2^E \to \mathcal{C}(A),$$

a measure of noncompactness on E, *and* $K \subset E$ *a closed pointed convex cone. Let* $f : K \to E$ *be a* k-Ψ-*contraction* ($k : \mathcal{A} \to [0,1]$). *If* f *is HLAL along* K, *then* $\mathcal{D}_\infty f|_K$ *is a* k-Ψ-*contraction.*

Proof. Let $\alpha \in \mathcal{A}$. From Property (7) cited in Definition 2.18, there is a neighbourhood U_α of zero in E such that $\Psi(U_\alpha)(\alpha) = 1$. Choose $\beta \in \mathcal{A}$ so that $B_\beta \subseteq U_\alpha$. Because f is HLAL along K, we can find $\gamma \in \mathcal{A}$ so that

(i) $K \setminus \ker p_\gamma \neq \emptyset$ (because E is Hausdorff).

(ii) For every $\varepsilon > 0$ there is $M > 0$ such that

$$p_\beta(f(x) - \mathcal{D}_\infty f(x)) \leq \varepsilon p_\gamma(x),$$

whenever $p_\gamma(x) \geq M$ and $x \in K$.

Let $S \subset K$ be bounded, and suppose for the moment that $p_\gamma \geq 1$ for every $x \in S$. Let $\varepsilon > 0$ be given, and put

$$\sigma = \sup\{p_\gamma(x) : x \in S\}.$$

Find $M > 0$ so that we have

$$p_\beta(f(x) - \mathcal{D}_\infty f(x)) \leq \frac{\varepsilon}{\sigma} p_\gamma(x),$$

for every $x \in K$ with $p_\gamma(x) \geq M$. Let λ be a real number such that $\lambda \geq M$. Then, if $x \in \lambda S$ (i.e., $x = \lambda x_s$ for some $x_s \in S$), we have

$$p_\gamma(x) = p_\gamma(\lambda x_s) = \lambda p_\gamma(x_s) \geq M p_\gamma(x_s) \geq M,$$

and therefore

$$p_\beta(f(x) - \mathcal{D}_\infty f(x)) \leq \frac{\varepsilon}{\sigma} p_\gamma(x) \leq \frac{\varepsilon}{\sigma} \lambda p_\gamma(x_s) \leq \varepsilon \lambda.$$

We deduce that

$$f(\lambda S) - \mathcal{D}_\infty f(\lambda S) \subseteq \varepsilon \lambda B_\beta \subseteq \varepsilon \lambda U_\alpha$$

and it follows that

$$\mathcal{D}_\infty f(\lambda S) \subseteq f(\lambda S) - [f(\lambda S) - \mathcal{D}_\infty f(\lambda S)] \subseteq f(\lambda S) + \varepsilon \lambda U_\alpha,$$

which implies

$$\lambda \Psi(\mathcal{D}_\infty f(S))(\alpha) = \Psi(\mathcal{D}_\infty f(\lambda S))(\alpha) \leq \Psi(f(\lambda S) + \varepsilon \lambda U_\alpha)(\alpha)$$
$$\leq \Psi(f(\lambda S))(\alpha) + \Psi(\varepsilon \lambda U_\alpha)(\alpha) \leq k(\alpha) \Psi(\lambda S)(\alpha) + \varepsilon \lambda$$
$$= \lambda(k(\alpha) \Psi(S)(\alpha) + \varepsilon).$$

Dividing through by λ and letting ε go to zero, we obtain

$$\Psi(\mathcal{D}_\infty f(S))(\alpha) \leq k(\alpha)\Psi(S)(\alpha).$$

Now, if S is an arbitrary bounded subset of K, let

$$\delta = 1 + \sup\{p_\alpha(x) : x \in S\}.$$

Because $K \setminus \ker p_\gamma \neq \emptyset$, we can find $x_0 \in K$ such that $p_\gamma(x_0) = \delta$. Considering $S' = x_0 + S$, we find for $x = x_0 + x_s \in S'$ that

$$p_\gamma(x) = p_\gamma(x_0 + x_s) \geq p_\gamma(x_0) - p_\gamma(x_s) = \delta - p_\gamma(x_s)$$
$$\geq \delta - (\delta - 1) = 1.$$

But we have just seen above that

$$\Psi(\mathcal{D}_\infty f(S'))(\alpha) \leq k(\alpha)\Psi(S')(\alpha).$$

We have

$$\Psi(\mathcal{D}_\infty f(S'))(\alpha) = \Psi(\mathcal{D}_\infty f(x_0 + S))(\alpha) = \Psi(\mathcal{D}_\infty f(x_0) + \mathcal{D}_\infty f(S))(\alpha)$$
$$= \Psi(\mathcal{D}_\infty f(S))(\alpha),$$

and

$$\Psi(S')(\alpha) = \Psi(x_0 + S)(\alpha) = \Psi(S)(\alpha),$$

which imply that the proof is complete. □

We have a similar result for the $\mathcal{B}$-asymptotic linearity.

THEOREM 2.25 *Let $(E, \{p_\alpha\}_{\alpha \in A})$ be a Hausdorff locally convex space*

$$\Psi : 2^E \to \mathcal{C}(A)$$

a measure of noncompactness on E and $K \subset E$ a closed pointed convex cone. Let $f : K \to E$ be a k-Ψ-contraction ($k : A \to [0,1]$). If f is $\mathcal{B}$AL along K for some basis $\mathcal{B}$ of the bornology on E, then $\mathbf{D}_\infty|_K$ is a k-Ψ-contraction.

Proof. Let $\alpha \in A$ and let U_α be a convex neighbourhood of zero with $\Psi(U_\alpha)(\alpha) = 1$. Select $\beta \in A$ so that $B_\beta \subset U_\alpha$. If $S \subset K$ is bounded, find $B \in \mathcal{B}$ such that $S \subset E_B$ and $S \subset \mu B$ for some $\mu > 0$. If $\varepsilon > 0$ is given, we assume for the moment that $p_B(x) \geq 1$ for every $x \in S$. We then find $M > 0$ so that $x \in K$ and $p_B(x) \leq M$ imply that

$$p_B(f(x) - \mathbf{D}_\infty f(x)) \leq \varepsilon p_B(x).$$

Let λ be a real number such that $\lambda \geq M$. When $x \in \lambda S$, we have $p_B(x) \geq M$ so that

$$p_B(f(x) - \mathbf{D}_\infty f(x)) \leq \varepsilon p_B(x).$$

Hence, we have
$$f(\lambda S) - \mathbf{D}_\infty f(\lambda S) \subset \varepsilon \lambda B_\beta,$$
which implies that

$$\mathbf{D}_\infty f(\lambda S) \subseteq f(\lambda S) - [f(\lambda S) - D_\infty f(\lambda S)] \subseteq f(\lambda S) + \varepsilon \lambda B_\beta.$$

Thus, exactly as in the previous argument, we have

$$\Psi(\mathbf{D}_\infty f(S))(\alpha) \le k(\alpha)\Psi(S)(\alpha).$$

Now, if $S \subset K$ is an arbitrary bounded set, note that we may assume that $E_B \cap K \neq \{0\}$ (we need only to choose $x_0 \in K$, $x_0 \neq 0$, and require $S \cup \{x_0\} \subset E_B$, rather than $S \subset E_B$). Moreover, $S \subset \mu B$ for some $\mu > 0$ means that S is bounded for the norm p_B on E_B. We can repeat the translation argument used in the proof of Theorem 2.24, using p_B in place of p_γ. The proof is complete. $\square$

REMARK 2.9 *The condition that f is a k-Ψ-contraction in Theorem 2.24 and in Theorem 2.25 cannot be relaxed to the condition that f is Ψ-condensing.*

Now we cite, without proof, a fixed point theorem in locally convex spaces due to Talman which is based on the notion of an asymptotic derivative along a cone.

THEOREM 2.26 *Let $(E, \{p_\alpha\}_{\alpha \in \mathcal{A}})$ be a Hausdorff, quasi-complete locally convex space. Let $K \subset E$ be a sharp total positive cone and let $f : K \to K$ be a continuous k-Ψ-contraction which is HL-asymptotically linear (respectively, $\mathcal{B}$-asymptotically linear for some basis $\mathcal{B}$ for the bornology on E). If $\mathcal{D}_\infty f$ (respectively, $\mathbf{D}_\infty f$) does not have any positive eigenvector belonging to an eigenvalue which is greater than or equal to one, then f has a fixed point in K.*

Proof. A proof of this theorem is in Talman [1973] and it is based on several intermediate results and on the topological index. $\square$

REMARK 2.10 *We note that Theorem 2.26 is a generalization of Krasnoselskii's fixed point theorem.*

2.5 The Asymptotic Scalar Differentiability

Inspired by the notion of scalar derivatives Isac introduced in 1999 the notion of the asymptotic scalar derivative [Isac, 1999c]. In this section we present this notion and some relations with the scalar derivative.

Let $(E, \|\cdot\|)$ be an arbitrary real Banach space. We say that a *semi-inner product* (in Lumer's sense) is defined on E, if to any $x, y \in E$ there corresponds a real number denoted by $[x, y]$ satisfying the following properties.

(s_1) $[x + y, z] = [x, z] + [y, z]$.

(s_2) $[\lambda x, y] = \lambda[x, y]$, for $x, y, z \in E$, $\lambda \in \mathbb{R}$.

(s_3) $[x, x] > 0$ for $x \neq 0$.

(s_4) $|[x, y]|^2 \leq [x, x][y, y]$.

It is known [Giles, 1967; Lumer, 1961] that a semi-inner product space is a normed linear space with the norm $\|x\|_s = [x, x]^{1/2}$ and that every Banach space can be endowed with a semi-inner product (and in general in infinitely many different ways, but a Hilbert space in a unique way).

Obviously if $(H, \langle\cdot, \cdot\rangle)$ is a Hilbert space, the inner product $\langle\cdot, \cdot\rangle$ is the unique semi-inner product in Lumer's sense on H, [Giles, 1967; Lumer, 1961].

We note that it is possible to define a semi-inner product such that $[x, x] = \|x\|^2$ (where $\|\cdot\|$ is the norm given in E). In this case we say that the semi-inner product is *compatible with the norm* $\|\cdot\|$. By the proof of Theorem 1 [Giles, 1967] this semi-inner product can be defined to have the homogeneity property:

(s_5) $[x, \lambda y] = \lambda[x, y]$, for $x, y \in E$, $\lambda \in \mathbb{R}$.

Throughout this chapter we suppose that all semi-inner products compatible with the norm satisfy (s_5).

The following definition is an extension of Example 5.1, p.169 of [do Carmo, 1992].

DEFINITION 2.27 *The mapping*

$$i : E\backslash\{0\} \to E\backslash\{0\}; \ i(x) = \frac{x}{[x, x]}$$

is called the inversion (of pole 0) *with respect to* $[\cdot, \cdot]$.

It is easy to see that i is one to one and $i^{-1} = i$. Indeed, because

$$\|i(x)\|_s = \frac{1}{\|x\|_s},$$

by the definition of i we have

$$i(i(x)) = \frac{i(x)}{\|i(x)\|_s^2} = \|x\|_s^2 i(x) = x.$$

Hence i is a global homeomorphism of $E\backslash\{0\}$ which can be viewed as a global nonlinear coordinate transformation in E.

Let $A \subseteq E$ such that $0 \in A$ and $A\backslash\{0\}$ is an invariant set of the inversion i with respect to $[\cdot, \cdot]$; that is, $i(A\backslash\{0\}) = A\backslash\{0\}$ and $f : A \to E$. Examples of invariant sets of the inversion i with respect to $[\cdot, \cdot]$ are:

1. $F \backslash \{0\}$ where F is a linear subspace of E (in particular F can be the whole E)

2. $K \backslash \{0\}$ where $K \subseteq E$ is a convex cone

Now we define the inversion (of pole 0) with respect to $[\cdot, \cdot]$ of the mapping f.

DEFINITION 2.28 *The* inversion (of pole 0) *with respect to* $[\cdot, \cdot]$ *of the mapping* f *is the mapping* $\mathcal{I}(f) : A \to E$ *defined by:*

$$\mathcal{I}(f)(x) = \begin{cases} [x, x](f \circ i)(x) & \text{if} \quad x \neq 0, \\ 0 & \text{if} \quad x = 0. \end{cases}$$

PROPOSITION 2.9 *The inversion of mappings $\mathcal{I}$ with respect to $[\cdot, \cdot]$ is a one-to-one mapping on the set of mappings $\{f \mid f : A \to E; f(0) = 0\}$ and $\mathcal{I}^{-1} = \mathcal{I}$; that is, $\mathcal{I}(\mathcal{I}(f)) = f$.*

Proof. By definition $\mathcal{I}(\mathcal{I}(f))(0) = 0$. Hence, $\mathcal{I}(\mathcal{I}(f))(0) = f(0)$. If $x \neq 0$, then $\mathcal{I}(\mathcal{I}(f))(x) = \|x\|_s^2 \mathcal{I}(f)(i(x)) = \|x\|_s^2 \|i(x)\|_s^2 f(i(i(x))) = f(x)$. Thus, $\mathcal{I}(\mathcal{I}(f))(x) = f(x)$ for all $x \in A$. Therefore, $\mathcal{I}(\mathcal{I}(f)) = f$. $\square$

REMARK 2.11 *We note that the inversion of mappings with respect to $[\cdot, \cdot]$ is linear and has the following properties.*

1. *If $T \in L(E, E)$ and $j : A \hookrightarrow E$ is the embedding of A into E, then $\mathcal{I}(T \circ j) = T \circ j$.*

2. *If the semi-inner product is compatible with the norm of E and $\|x\| \to +\infty$, then $i(x) \to 0$.*

Now, we introduce the notion of a scalar derivative with respect to a semi-inner product $[\cdot, \cdot]$.

Let $(E, \| \cdot \|)$ be an arbitrary real Banach space and $[\cdot, \cdot]$ a semi-inner product on E. Let $G \subseteq E$ be a set which contains at least one nonisolated point, $\widetilde{G} \subseteq E$ such that $G \subseteq \widetilde{G}$, $f : \widetilde{G} \to E$ and x_0 a nonisolated point of G. The following definition is an extension of Definition 2.2 [Nemeth, 1992].

DEFINITION 2.29 *The limit*

$$\underline{f}^{\#,G}(x_0) = \liminf_{\substack{x \to x_0 \\ x \in G}} \frac{[f(x) - f(x_0), x - x_0]}{\|x - x_0\|_s^2}$$

is called the lower scalar derivative *of f at x_0 along G with respect to $[\cdot, \cdot]$. Taking* $\limsup$ *in place of* $\liminf$, *we can define the* upper scalar derivative $\overline{f}^{\#,G}(x_0)$ *of f at x_0 along G with respect to $[\cdot, \cdot]$ similarly.*

REMARK 2.12 *If $G = \widetilde{G}$, then without confusion, we can say, for short, lower scalar derivative and upper scalar derivative instead of lower scalar derivative along G and upper scalar derivative along G, respectively. In this case, we omit G from the superscript of the corresponding notations.*

PROPOSITION 2.10 *Suppose that $[\cdot, \cdot]$ is compatible with the norm $\| \cdot \|$. Let $K \subseteq E$ be an unbounded set such that $0 \in K$ and $K \backslash \{0\}$ is an invariant set of the inversion i with respect to $[\cdot, \cdot]$. Let $g : E \to E$. Then we have*

$$\liminf_{\substack{\|x\| \to \infty \\ x \in K}} \frac{[g(x), x]}{\|x\|^2} = \mathcal{I}(g)^{\#,K}(0).$$

Proof. Because $K \subseteq E$ is unbounded and $K \backslash \{0\}$ is an invariant set of i, 0 is a nonisolated point of K. Hence, $\mathcal{I}(g)^{\#,K}(0)$ is well defined. Consider the global nonlinear coordinate transformation $y = i(x)$. Then $x = i(y)$ and we have

$$\liminf_{\substack{\|x\| \to \infty \\ x \in K}} \frac{[g(x), x]}{\|x\|^2} = \liminf_{\substack{y \to 0 \\ y \in K}} [\mathcal{I}(g)(y), i(y)],$$

from where, by using the definition of the lower scalar derivative along a set, the assertion of the lemma follows easily. $\qquad \square$

REMARK 2.13 *Obviously, if the Banach space $(E, \| \cdot \|)$ is a Hilbert space $(H, \langle \cdot, \cdot \rangle)$, in Definition 2.29 and Proposition 2.10 we replace the semi-inner product $[\cdot, \cdot]$ by the inner product $\langle \cdot, \cdot \rangle$ defined on H.*

Let $(E, \| \cdot \|)$ be an arbitrary Banach space, $[\cdot, \cdot]$ a semi-inner product on E, and $K \subset E$ an unbounded set .

The following definition is an extension of the notion of an asymptotic scalar derivative given on Hilbert space by Isac [Isac, 1999c].

Let $f : K \to E$ be an arbitrary mapping.

DEFINITION 2.30 *We say that $T \in L(E, E)$ is an asymptotic scalar derivative of f along K, with respect to the semi-inner product $[\cdot, \cdot]$ if*

$$\limsup_{\substack{\|x\| \to \infty \\ x \in K}} \frac{[f(x) - T(x), x]}{\|x\|_s^2} \leq 0.$$

The mapping of Definition 2.30 is denoted $f'_{s,K}(\infty)$. For the next results we suppose that $0 \in K$ and $K \backslash \{0\}$ is an invariant set of the inversion i with respect to $[\cdot, \cdot]$.

REMARK 2.14 *If the semi-inner product $[\cdot, \cdot]$ is compatible with the norm $\| \cdot \|$, then in Definitions 2.29 and 2.30 we can replace $\|x - x_0\|_s^2$ by $\|x - x_0\|^2$ and $\|x\|_s^2$ by $\|x\|^2$, respectively.*

PROPOSITION 2.11 *If T is an asymptotic scalar derivative of f with respect to the semi-inner product $[\cdot, \cdot]$, then for any $c > 0$ the mapping $T + cI$ is also an asymptotic scalar derivative of f with respect to $[\cdot, \cdot]$.*

Proof. This proposition is a consequence of Definition 2.30.

THEOREM 2.31 *If $[\cdot, \cdot]$ is a semi-inner product compatible with the norm $\| \cdot \|$, then $T \in L(E)$ is an asymptotic scalar derivative of f with respect to $[\cdot, \cdot]$ if and only if the upper scalar derivative of h in 0 is nonpositive (i.e., $\overline{h}^{\#}(0) \leq 0$), where $h : K \to E$ is defined by $h = \mathcal{I}(f - T \circ j) = \mathcal{I}(f) - T \circ j$, and $j : K \hookrightarrow E$ is the embedding of K into E.*

Proof. We suppose that $T \in L(E)$ is an asymptotic scalar derivative of f with respect to the semi-inner product $[\cdot, \cdot]$ and prove that $\overline{h}^{\#}(0) \leq 0$. The converse implication can be proved similarly. Indeed, because $T \in L(E)$ is an asymptotic scalar derivative of f with respect to $[\cdot, \cdot]$, we have that

$$\limsup_{\substack{\|x\| \to +\infty \\ x \in K}} [f(x) - T(x), i(x)] \leq 0. \tag{2.12}$$

Consider the global nonlinear coordinate transformation $y = i(x)$ given by the global diffeomorphism i. Because K is unbounded and $K \backslash \{0\}$ is invariant under i, 0 is a nonisolated point of K. Then, $x = i(y)$ and by (2.12),

$$\limsup_{\substack{y \to 0 \\ y \in K}} [(f \circ i)(y) - (T \circ j \circ i)(y), y] \leq 0.$$

Hence,

$$\limsup_{\substack{y \to 0 \\ y \in K}} [\mathcal{I}(f)(y) - \mathcal{I}(T \circ j)(y), i(y)] \leq 0.$$

Thus, by the definition of the upper scalar derivative with respect to $[\cdot, \cdot]$ we have $\overline{h}^{\#}(0) \leq 0$.

COROLLARY 2.32 *If the semi-inner product $[\cdot, \cdot]$ is compatible with the norm $\| \cdot \|$, then 0 is an asymptotic scalar derivative of f with respect to $[\cdot, \cdot]$ if and only if $\overline{\mathcal{I}(f)}^{\#}(0) \leq 0$.*

The following theorem shows the surprising fact that if $[\cdot, \cdot]$ is compatible with the norm $\| \cdot \|$, then every f whose inversion has a finite upper scalar derivative with respect to $[\cdot, \cdot]$ at 0 is asymptotically scalarly differentiable with respect to $[\cdot, \cdot]$.

THEOREM 2.33 *If $[\cdot, \cdot]$ is a semi-inner product compatible with the norm $\| \cdot \|$ and $\overline{\mathcal{I}(f)}^{\#}(0) < +\infty$, then f is asymptotically scalarly differentiable with* respect to $[\cdot, \cdot]$ and $T = \overline{\mathcal{I}(f)}^{\#}(0)I$ is an asymptotic scalar derivative of f with respect to $[\cdot, \cdot]$, where $I : E \to E$ is the identity mapping (the asymptotic scalar differentiability is along the unbounded set K).

Proof. Indeed, $\overline{h}^{\#}(0) = 0$, where $h = \mathcal{I}(f) - T \circ j = \mathcal{I}(f) - \overline{\mathcal{I}(f)}^{\#}(0)(I \circ j)$. Hence, the result follows by using Theorem 2.31.

REMARK 2.15 *If the semi-inner product $[\cdot, \cdot]$ is compatible with the norm $\| \cdot \|$ and $\overline{\mathcal{I}(f)}^{\#}(0) < +\infty$, then every mapping cI is an asymptotic scalar derivative with respect to $[\cdot, \cdot]$, where $c \geq \overline{\mathcal{I}(f)}^{\#}(0)$.*

If the Banach space $(E, \| \cdot \|)$ is in particular a Hilbert space and the norm $\| \cdot \|$ is the norm defined by the inner product $\langle \cdot, \cdot \rangle$ given on the vector space H, then Definition 2.30 has the following form.

DEFINITION 2.34 *Let $(H, \langle \cdot, \cdot \rangle)$ be a Hilbert space and $K \subset H$ an unbounded set. We say that $T \in L(H, H)$ is an asymptotic scalar derivative of $f : K \to H$, along K if*

$$\limsup_{\substack{\|x\| \to \infty \\ x \in K}} \frac{\langle f(x) - T(x), x \rangle}{\|x\|^2} \leq 0.$$

Now, we consider a more general situation. Let $(E, \| \cdot \|)$ be a Banach space, E^* the topological dual of E, $\langle E, E^* \rangle$ a duality between E and E^* with respect to a bilinear functional on $E \times E^*$, denoted $\langle \cdot, \cdot \rangle$ and satisfying the separation axioms. Let $K \subseteq E$ be an unbounded set $\tilde{K} \subset E$ such that $K \subseteq \tilde{K}$ and $f : \tilde{K} \to E^*$ be a mapping.

DEFINITION 2.35 *We say that $T \in L(E, E^*)$ is an asymptotic scalar derivative of f along K if*

$$\limsup_{\substack{\|x\| \to +\infty \\ x \in K}} \frac{\langle x, f(x) - T(x) \rangle}{\|x\|^2} \leq 0.$$

The mapping used in Definition 2.35 is denoted $f'_{s,K}(\infty)$. If $K = \tilde{K}$ we can say asymptotic scalar derivative for short instead of asymptotic scalar derivative along K.

REMARK 2.16 *If in Definitions 2.30, 2.34, and 2.35 we have that $K = E$, $K = H$, and $K = E$, respectively, we say that T is the asymptotic scalar derivative with respect to the space E, H, E, respectively.*

Let $(E, \| \cdot \|)$ be a Banach space and $[\cdot, \cdot]$ a semi-inner product (in Lumer's sense) and let $\| \cdot \|_s$ be the norm defined by this semi-inner product. Let $K \subset E$ be a closed convex cone and $f : E \to E$.

DEFINITION 2.36 *We say that $T \in L(E, E)$ is an asymptotic derivative of f along K if*

$$\lim_{\substack{\|x\|_s \to +\infty \\ x \in K}} \frac{\|f(x) - T(x)\|_s}{\|x\|_s} = 0.$$

PROPOSITION 2.12 *If $T \in L(E, E)$ is an asymptotic derivative of f along K, then T is an asymptotic scalar derivative along K.*

Proof. The proposition is a consequence of the relation

$$\limsup_{\substack{\|x\|_s \to +\infty \\ x \in K}} \frac{[f(x) - T(x), x]}{\|x\|_s^2} \leq \limsup_{\substack{\|x\|_s \to +\infty \\ x \in K}} \frac{\|f(x) - T(x)\|_s \|x\|_s}{\|x\|_s^2}$$

$$= \lim_{\substack{\|x\|_s \to +\infty \\ x \in K}} \frac{\|f(x) - T(x)\|_s}{\|x\|_s} = 0.$$

$\square$

REMARK 2.17 *Let $(H, \langle \cdot, \cdot \rangle)$ be a Hilbert space. By the definition of the asymptotic scalar derivative, it follows easily that if U is an asymptotic scalar derivative of f and $g : H \to H$ satisfies the relation*

$$\langle g(x), x \rangle \leq 0, \tag{2.13}$$

for all $x \in H$, then U is also an asymptotic scalar derivative of $f + g$. Particularly, for any skew-adjoint mapping Z, the mapping U is an asymptotic scalar derivative of $f + Z$, or equivalently $U + Z$ is an asymptotic scalar derivative of f. Moreover, for any P continuous linear positive semi-definite mapping, $U + P$ is also an asymptotic scalar derivative of f. An example for a nonlinear mapping g satisfying (2.13) is $g : R^3 \to R^3$;

$$g(u, v, w) = (-u + vw, -v + uw, -w - 2uv).$$

It would be interesting to study the properties of mappings satisfying the condition (2.13). Of course, 0 is an asymptotic scalar derivative of these mappings.

REMARK 2.18 *We have already shown that every asymptotic derivative of f is an asymptotic scalar derivative of f. However, the converse is not true. Indeed, it can be easily checked that if $f : R^3 \to R^3$,*

$$f(u, v, w) = (vw, uw, -2uv),$$

then 0 is an asymptotic scalar derivative of f but it is not an asymptotic derivative of f.

REMARK 2.19 *Every continuous mapping S satisfying (2.7) is an asymptotic scalar derivative of f. Indeed, we have*

$$\limsup_{\|x\|\to+\infty} \frac{\langle f(x) - T(x), x \rangle}{\|x\|^2} \leq \frac{2\theta}{2 - \psi(2)} \lim_{\|x\|\to+\infty} \frac{\psi(\|x\|)}{\|x\|} = 0.$$

Let $K \subset E$ be a closed convex cone and $f : K \to E$ a mapping. Let $\mathcal{R} : H \to K$ be a retraction. We have the following result.

PROPOSITION 2.13 *If the retraction $\mathcal{R}$ is a ρ-Lipschitz mapping with respect to the norm $\| \cdot \|_s$ and $T \in L(E, E)$ is an asymptotic derivative of f along K such that $T(K) \subseteq K$, then T is an asymptotic scalar derivative of $\mathcal{R} \circ f$ along K.*

Proof. Indeed, we have

$$\lim_{\substack{\|x\|_s\to+\infty \\ x\in K}} \frac{[\mathcal{R}(f(x)) - T(x), x]}{\|x\|_s^2} = \lim_{\substack{\|x\|_s\to+\infty \\ x\in K}} \frac{[\mathcal{R}(f(x)) - \mathcal{R}(T(x)), x]}{\|x\|_s^2}$$

$$\leq \lim_{\substack{\|x\|_s\to+\infty \\ x\in K}} \frac{\|\mathcal{R}(f(x)) - \mathcal{R}(T(x))\|_s\|x\|_s}{\|x\|_s^2} \leq \lim_{\substack{\|x\|_s\to+\infty \\ x\in K}} \frac{\rho\|f(x) - T(x)\|_s\|x\|_s}{\|x\|_s^2}$$

$$= 0.$$

$\square$

REMARK 2.20 *We note that Proposition 2.13 has interesting applications to the study of nonlinear complementarity problems in Hilbert spaces. In this case the retraction $\mathcal{R}$ is the projection onto a closed convex cone.*

2.6   Some Applications

We present in this section some applications to the study of fixed points of nonlinear mappings and also to the study of nonlinear complementarity problems.

First, we give an interesting variant of Krasnoselskii's fixed point theorem (Theorem 2.7). We need to introduce a notation.

Let $(H, \langle \cdot, \cdot \rangle)$ be a Hilbert space, $K \subset H$ a closed convex cone, K^* its dual cone, and $f, g : K \to H$ two mappings. The relation $h \leq_{K^*} g$ means that $g(x) - f(x) \in K^*$ (the dual of the cone K), for all $x \in K$. In this case we have in particular $\langle h(x), x \rangle \leq \langle g(x), x \rangle$ for all $x \in K$.

DEFINITION 2.37 *We say that a mapping $f : K \to H$ is scalarly compact, if for any sequence $\{x_n\}_{n\in\mathbb{N}} \subset K$ weakly convergent to an element $x_* \in K$, there exists a subsequence $\{x_{n_k}\}_{k\in\mathbb{N}}$ such that*

$$\limsup_{k\to+\infty} \langle x_{n_k} - x_*, f(x_{n_k}) \rangle \leq 0.$$

Examples

(1) Any completely continuous mapping is scalarly compact.

(2) Given a mapping $f : K \to H$, if there exists a completely continuous mapping $h : K \to H$ such that

$$\langle y, f(x) \rangle \leq |\langle y, h(x) \rangle|$$

for any $x, y \in K$, then f is scalarly compact .

THEOREM 2.38 *Let $(H, \langle \cdot, \cdot \rangle)$ be a Hilbert space, $K \subset H$ a pointed closed convex cone, and $f : K \to K$ a mapping.*
If the following assumptions are satisfied,

(i) *f is demicontinuous;*

(ii) *f is scalarly compact;*

(iii) *there exists an asymptotic scalarly differentiable mapping $f_0 : K \to H$ such that $f \leq_{K^*} f_0$ and $\|f'_{0s}(\infty)\| < 1$;*

then f has a fixed point in K.

Proof. We use the notion of a nonlinear complementarity problem defined in (2.21) and the notion of a variational inequality defined in (3.9). We define $h = I - f$, where I is the identity mapping. From the complementarity theory we know that f has a fixed point in K if and only if the nonlinear complementarity problem NCP(h, K) has a solution.

For every $m \in \mathbb{N}$ we define the set

$$K_m = \{x \in K : \|x\| \leq m\}$$

and we observe that K_m is closed, convex, weakly closed, and $K = \cup_{m=1}^{\infty} K_m$. Obviously, any set K_m is bounded.

First, we show that for every $m \in \mathbb{N}$ the variational inequality VI$(I - f, K_m)$ has a solution $y_m^* \in K_m$. Indeed, let $m \in N$ arbitrary and denote by Λ the family of all finite-dimensional subspaces of H ordered by inclusion. Consider the mapping $h(x) = x - f(x)$ for all $x \in K$ and define $K_m(E) = K_m \cap E$ for each $E \in \Lambda$.

For each $E \in \Lambda$ we set

$$A_E = \{y \in K_m : \langle h(y), x - y \rangle \geq 0 \text{ for all } x \in K_m(E)\}$$

and we have that A_E is nonempty. Indeed, the solution set of the problem VI$(h, K_m(E))$ is a subset of A_E, but the solution set of VI$(h, K_m(E))$ is nonempty. To see this, we consider the mappings $j : E \to H$ and $j^* : H^* \to E^*$, where j is the inclusion and j^* is the adjoint of j. The mapping $j^* \circ h \circ j :$ $K_m(E) \to E^*$ is continuous and

$$\langle j^* \circ h \circ j(y), x - y \rangle = \langle h(j(y)), j(x - y) \rangle = \langle h(y), x - y \rangle,$$

for all $x, y \in K_m(E)$. Applying the classical Hartman–Stampacchia theorem to the set $K_m(E)$ and to the mapping $j^* \circ h \circ j$ we obtain that the problem $VI(h, K_m(E))$ has at least a solution.

For every $E \subset \Lambda$ we denote $\bar{A}_E^\sigma$ the weak closure of A_E. We have that $\cap_{E \in \Lambda} \bar{A}_E^\sigma$ is nonempty. Indeed, let $\bar{A}_{E_1}^\sigma, \bar{A}_{E_2}^\sigma, \ldots, \bar{A}_{E_n}^\sigma$ be a finite subfamily of the family $\{\bar{A}_E^\sigma\}_E \in \Lambda$. Let M be the finite-dimensional subspace generated by $E_1, E_2, \ldots, E_n$.

Because $E_k \subseteq M$ for all $k \in \{1, 2, \ldots, n\}$, we have that $K_m(E_k) \subseteq E_m(M)$ for all $k \in \{1, 2, \ldots, n\}$. Therefore, $A_M \subseteq A_{E_K}$ for all $k \in \{1, 2, \ldots, n\}$, which implies that $\cap_{k=1} \bar{A}_{E_K}^\sigma$ is nonempty. The weak compactness of K_m implies that $\cap_{E \in \Lambda} \bar{A}_E^\sigma \neq \emptyset$. Let $y_m^* \in \cap_{E \in \Lambda} \bar{A}_E^\sigma$ be arbitrary and let $x \in K_m$ be any element of this set. There exists some $E \in \Lambda$ such that $x, y_m^* \in E$. Because $y_m^* \in \bar{A}_E^\sigma$, there exists a sequence $\{y_n\}_{n \in \mathbb{N}} \subset A_E$ such that $\{y_n\}_{n \in \mathbb{N}}$ is weakly convergent to y_m^* (we applied Šmulian's theorem). We have

$$\langle h(y_n), y_m^* - y_n \rangle \geq 0,$$

and

$$\langle h(y_n), x - y_n \rangle \geq 0,$$

or

$$\langle y_n, y_n - y_m^* \rangle \leq \langle f(y_n), y_n - y_m^* \rangle, \tag{2.14}$$

and

$$\langle y_n, x - y_n \rangle \leq \langle f(y_n), x - y_n \rangle. \tag{2.15}$$

From (2.14) and assumption (ii) we have that $\{y_n\}$ has a subsequence, denoted again $\{y_n\}$, such that

$$\limsup_{n \to \infty} \langle y_n, y_n - y_m^* \rangle \leq 0, \tag{2.16}$$

which implies

$$0 \leq \limsup_{n \to \infty} \|y_n - y_m^*\|^2 = \limsup_{n \to \infty} \langle y_n - y_m^*, y_n - y_m^* \rangle$$
$$\leq \limsup_{n \to \infty} \langle y_n, y_n - y_m^* \rangle + \limsup_{n \to \infty} [-\langle y_m^*, y_n - y_m^* \rangle] \leq 0.$$

We deduce that $\{y_n\}$ is strongly convergent to y_m^*. Because f is demicontinuous, we have that $\{f(y_n)\}_{n \in \mathbb{N}}$ is weakly convergent to $f(y_m^*)$.

From (2.15) we have

$$\langle y_m^* - f(y_m^*), x - y_m^* \rangle \geq 0$$

for any $x \in K_m$; that is, y_m^* is a solution of VI($I - f, K_m$). (We note that to obtain the last inequality we also used the following fact, "If $\{u_n\}$ is weakly convergent to an element u_* and $\{v_n\}$ is strongly convergent to an element v_*, then $\lim_{n \to \infty} \langle u_n, v_n \rangle = \langle u_*, v_* \rangle$.")

Now, we pass to the second part of the proof. In the first part we proved that for every $m \in \mathbb{N}$ the problem VI($I - f, K_m$) has a solution y_m; that is,

$$\langle y_m - f(y_m), x - y_m \rangle \geq 0, \text{ for all } x \in K_m. \tag{2.17}$$

Taking $x = 0$ in (2.17), we obtain

$$\langle y_m, y_m \rangle \leq \langle f(y_m), y_m \rangle. \tag{2.18}$$

The sequence $\{y_m\}_{m \in \mathbb{N}}$ is bounded. Indeed, if this is false, we may assume that $\|y_m\| \to +\infty$ as $m \to +\infty$, which implies (using (2.18) and assumption (iii))

$$1 = \frac{\langle y_m, y_m \rangle}{\|y_m\|^2} \leq \lim_{\|y_m\| \to +\infty} \frac{\langle f(y_m), y_m \rangle}{\|y_m\|^2}$$

$$\leq \limsup_{\|y_m\| \to +\infty} \frac{\langle f_0(y_m), y_m \rangle}{\|y_m\|^2}$$

$$\leq \limsup_{\|y_m\| \to +\infty} \frac{\langle f_0(y_m) - f_{0s}'(\infty)(y_m), y_m \rangle}{\|y_m\|^2} + \limsup_{\|y_m\| \to \infty} \frac{\langle f_{0s}'(\infty)(y_m), y_m \rangle}{\|y_m\|^2}$$

$$\leq \limsup_{\|y_m\|^2} \frac{\|f_{0s}'(\infty)\| \|y_m\|^2}{\|y_m\|^2} = \|f_{0s}'(\infty)\| < 1.$$

We have a contradiction and therefore $\{y_m\}_{m \in \mathbb{N}}$ is a bounded sequence. By the reflexivity of H and the weak closedness of K we have that there exists a subsequence $\{y_{m_k}\}_{k \in \mathbb{N}}$ of the sequence $\{y_m\}_{m \in \mathbb{N}}$, weakly convergent to $y_0 \in K$. For all $x \in K$, there exists an $m_0 \in \mathbb{N}$ such that y_0 and x are in K_{m_0}. Thus, for all $m \geq 0$ we have $y_0, x \in K_m$. We have

$$\langle y_m - f(y_m), y_0 - y_m \rangle \geq 0. \tag{2.19}$$

and

$$\langle y_m - f(y_m), x - y_m \rangle \geq 0 \tag{2.20}$$

Using inequality (2.19) and the scalar compactness of f (i.e., assumption (ii)) we have that there exists a subsequence $\{y_{m_k}\}_{k \in \mathbb{N}}$ of the sequence $\{y_m\}_{m \in \mathbb{N}}$ such that

$$\limsup_{k \to \infty} \langle y_{m_k}, y_{m_k} - y_0 \rangle \leq \limsup_{k \to \infty} \langle f(y_{m_k}), y_{m_k} - y_0 \rangle \leq 0$$

which implies that $\{y_{m_k}\}_{k \in \mathbb{N}}$ is strongly convergent to y_0, as we can see considering the following inequalities,

$$0 \leq \limsup_{k \to \infty} \|y_{m_k} - y_0\|^2 = \limsup_{k \to \infty} \langle y_{m_k} - y_0, y_{m_k} - y_0 \rangle$$

$$\leq \limsup_{k \to \infty} \langle y_{m_k}, y_{m_k} - y_0 \rangle + \limsup_{k \to \infty} [-\langle y_0, y_{m_k} - y_0 \rangle \leq 0.$$

Considering (2.20) for all $m_k \geq m_0$ we have

$$\langle y_{m_k} - f(y_{m_k}), x - y_{m_k} \rangle \geq 0.$$

Computing the limit in the last inequality we obtain

$$\langle y_0 - f(y_0), x - y_0 \rangle \geq 0$$

for any $x \in K$. Therefore, $f(y_0) = y_0$ and the proof is complete. $\square$

COROLLARY 2.39 *Let $(H, \langle \cdot, \cdot \rangle)$ be a Hilbert space, $K \subset H$ a pointed closed convex cone, and $f : K \to K$ a mapping. If the following assumptions are satisfied,*

(1) *f is demicontinuous;*

(2) *f is scalarly compact;*

(3) *f has an asymptotic scalar derivative $f'_s(\infty)$ and $\|f'_s(\infty)\| < 1$;*

then f has a fixed point in K

COROLLARY 2.40 *Let $(H, \langle \cdot, \cdot \rangle)$ be a Hilbert space and $K \subset H$ a generating closed pointed convex cone. Let $f : K \to K$ be a completely continuous mapping. If f is asymptotically linear and $\|f'(\infty)\| < 1$, then f has a fixed point in K.*

COROLLARY 2.41 *Let $(H, \langle \cdot, \cdot \rangle)$ be a Hilbert space and $K \subset H$ a generating closed pointed convex cone. Let $f : K \to K$ be a completely continuous mapping. If there exists an asymptotically linear mapping $f_0 : K \to K$ such that $f \leq_{K^*} f_0$ and $\|f'_s\| < 1$, then f has a fixed point in K.*

Let $(H, \langle \cdot, \cdot \rangle)$ be a Hilbert space, $K \subset H$ a closed pointed convex cone, K^* its dual cone, and $f : H \to H$ a mapping. We consider the general nonlinear complementarity problem

$$\text{NCP}(f, K) : \begin{cases} \text{find } x_0 \in K \text{ such that} \\ f(x_0) \in K^* \text{ and } \langle x_0, f(x_0) \rangle = 0. \end{cases} \quad (2.21)$$

We say that f is a *completely continuous field* if f has a representation of the form $f(x) = x - T(x)$, for any $x \in H$, where $T : H \to H$ is a completely continuous mapping . Also we say that f is an asymptotically differentiable field with respect to K if f has a representation of the form $f(x) = x - T(x)$, for any $x \in H$, where $T : H \to H$ has an asymptotic derivative T'_∞ along K. We have the following result related to the NCP(f, K) problem.

THEOREM 2.42 *Let $(H, \langle \cdot, \cdot \rangle)$ be a Hilbert space, $K \subset H$ a generating closed pointed convex cone, and $f : H \to H$ a mapping. The mapping f is supposed to be a completely continuous and asymptotically differentiable field of the form $f(x) = x - T(x)$ for any $x \in H$. If $\|T'_\infty\| < 1$ and $T'_\infty(K) \subseteq K$, then the problem NCP(f, K) has a solution.*

Proof. From the complementarity theory it is known that the problem NCP(f, K) has a solution if and only if the mapping

$$\Phi(x) = P_K(x - f(x)) = P_K(T(x))$$

has a fixed point. Obviously, $\Phi(K) \le K$ and Φ is a completely continuous mapping. Therefore, Φ is demicontinuous and scalarly compact. Because $T'_\infty(x) = P_K(T'_\infty(x))$, consequently

$$\lim_{\substack{\|x\| \to \infty \\ x \in K}} \frac{\|P_K(T(x)) - T'_\infty(x)\|}{\|x\|} \le \lim_{\substack{\|x\| \to \infty \\ x \in K}} \frac{\|P_K(T(x)) - P_K(T'_\infty(x))\|}{\|x\|}$$

$$\le \lim_{\substack{\|x\| \to \infty \\ x \in K}} \frac{\|T(x) - (T'_\infty(x))\|}{\|x\|} = 0.$$

We have that T'_∞ is also an asymptotic derivative of the mapping Φ, which implies that $\Phi'_s(\infty) = T'_\infty$. Because the assumptions of Theorem 2.38 are satisfied our theorem is proved. $\square$

Remarks

1. The assumption $T'_\infty(K) \subseteq K$ is satisfied if $T(K) \subseteq K$ (see [Krasnoselskii, 1964a]).

2. Theorem 2.42 is applicable to complementarity problems defined by completely continuous fields of the form $f(x) = x - T(x)$, where T is an integral operator. It is known that many nonlinear integral operators (as for example, Hammerstein operators or Urysohn mappings are asymptotically differentiable [Krasnoselskii, 1964b].

3. Theorem 2.42 is also applicable to complementarity problems NCP(f, K), where f has a representation of the form $f(x) = \alpha x - T(x)$, where α is a positive real number and $T : H \to H$ is a completely continuous mapping. In this case, in the proof of Theorem 2.42 we consider the mapping

$$\Psi(x) = \Phi_K \left(x - \frac{1}{\alpha} f(x) \right) = P_K \left(\frac{1}{\alpha} T(x) \right).$$

In this case we must ask to have $\|T'_\infty\| < \alpha$.

Another interesting application of Theorem 2.38 to complementarity problems is when the cone K is an isotone projection cone and K is self-adjoint; that is, $K = K^*$. In this case if $f(x) = x - T(x)$ and there exists a mapping T_0 such that $T_0 : H \to H$ and $T(x) \leq_K T_0(x)$ for any $x \in H$, we have that $P_K(T(x)) \leq_K P_K(T_0(x))$. If T_0 has an asymptotic derivative $(T_0)'_\infty$ such that $(T_0)'_\infty(K) \subseteq K$, and the mapping f is a completely continuous field, then by Theorem 2.38 we have that the problem $\text{NCP}(f, K)$ has a solution if in addition $\|(T_0)'_\infty\| < 1$.

Chapter 3

Scalar Derivatives in Hilbert Spaces

3.1 Calculus

3.1.1 Introduction

The behavior of the scalar product $\langle f(x) - f(y), x - y \rangle$ (with $(H, \langle \cdot, \cdot \rangle)$ a Hilbert space and $f : H \to H$) when x and y run over H is a good tool in characterizing important properties of f. In order to avoid the difficulties in considering multifunctions we only consider so-called *upper and lower scalar derivatives of rank k*, which are extensions of the Dini derivatives.

If f is bounded, then this product converges to 0 for $x \to y$. Therefore it cannot be used in obtaining a local characterization. Hence it is natural to consider at y limits of the expressions of form $\langle f(x) - f(y), x - y \rangle / \|x - y\|^p$ for $x \to y$, where $p > 0$ and $\| \cdot \|$ is the norm generated by the scalar product $\langle \cdot, \cdot \rangle$. Thus we arrive by a natural way to a notion which we call *scalar derivative of rank p*. For $p = 2$ we say, for short, *scalar derivative*. The case $p = 1$ is strongly related to the notion of submonotonicity (see [Georgiev, 1997] and [Spingarn, 1981]). The scalar derivative of rank p is in general a multivalued mapping from H to H even if f is a bounded linear mapping and $p = 2$. The case when upper and lower scalar derivatives coincide is considered only for $p = 2$.

In recent years scalar derivatives became an important tool in the study of fixed point theorems, surjectivity theorems, integral equations, variational inequalities, and complementarity problems [Isac and Nemeth, 2003, 2004, 2005a,b, 2006a,b, 2007a,b]. Therefore it is very important to give computational formulae for the scalar derivatives [Nemeth, 2006]. When scalar derivatives were introduced [Nemeth, 1992, 1993] the main purpose was to characterize monotone mappings. For simplification only the finite dimensional case was considered. This section intends to fill the gap between the computational

formulae given in Nemeth, 1993 and the formulae needed for computing types
of scalar derivatives which occur in [Isac and Nemeth, 2003, 2004, 2005a,b,
2006a,b, 2007a,b] when additional differentiability conditions are introduced.
The section is divided into six subsections. In Section 3.1.2 we present some
basic results concerning skew-adjoint operators. In Section 3.1.3 we introduce
the notion of scalar derivatives and show that scalar differentiability is a very
restrictive condition. In Section 3.1.4 we extend the correspondence between
scalar derivatives and monotone mappings from Euclidean spaces to Hilbert
spaces. In Section 3.1.5 we give new computational formulae for the scalar
derivatives. Many of these formulae arise from applications, such as fixed
point theorems, surjectivity theorems, integral equations, variational inequalities, and complementarity problems [Isac and Nemeth, 2003, 2004, 2005a,b,
2006a,b, 2007a,b]

3.1.2   Some Basic Results Concerning Skew-Adjoint Operators

From now on let $(H, \langle \cdot, \cdot \rangle)$ denote a Hilbert space and $\| \cdot \|$ the norm generated
by $\langle \cdot, \cdot \rangle$. We recall the following definition [Kachurovskii, 1960; Minty, 1962].

DEFINITION 3.1 *Consider the operator* $f : H \to H$. *It is called* monotone
if for any x *and* y *in* H *one has*

$$\langle f(x) - f(y), x - y \rangle \geq 0.$$

If

$$\langle f(x) - f(y), x - y \rangle > 0$$

whenever $x \neq y$, *then* f *is called* strictly monotone.

DEFINITION 3.2 *The linear operator* $A : H \to H$ *is* skew-adjoint *if for any*
x *and* y *in* H *the relation* $\langle Ax, y \rangle + \langle Ay, x \rangle = 0$ *holds.*

THEOREM 3.3 *If* $A : H \to H$ *is linear, then the following statements are
equivalent.*

1. *A is skew-adjoint.*

2. $\langle Ax, x \rangle = 0$ *for any* $x \in H$.

   Proof. $1 \Rightarrow 2$ Take x arbitrarily in H. By the definition of the skew-adjoint
operator A we have $\langle Ax, y \rangle + \langle Ay, x \rangle = 0$. Put $y = x$. Then $\langle Ax, x \rangle = 0$ for
arbitrary x in H. The implication $2 \Rightarrow 1$ can be shown similarly.   □

THEOREM 3.4 *Consider the operator $F : H \to H$. The following assertions are equivalent.*

1. $\langle F(x) - F(y), x - y \rangle = 0, \forall x, y \in H$ *(i.e., both F and $-F$ are monotone).*

2. *F is an affine operator with skew-adjoint linear term.*

Proof. Suppose that 1 holds. Put $f(x) = F(x) - F(0)$ for x in H. Then $f(0) = 0$ and $\langle f(x) - f(y), x - y \rangle = 0, \forall x, y \in H$. Let x be arbitrary in H and $y = 0$. Then $\langle f(x), x \rangle = 0, \forall x \in H$. The above relation also yields

$$\langle f(x), x \rangle - \langle f(x), y \rangle - \langle f(y), x \rangle + \langle f(y), y \rangle = 0, \forall x, y \in H$$

and hence

$$\langle f(x), y \rangle + \langle f(y), x \rangle = 0, \forall x, y \in H.$$

Put $x = \lambda x_1 + \mu x_2$ with arbitrary x_1 and x_2 in H. Then,

$$\langle f(\lambda x_1 + \mu x_2), y \rangle = -\langle f(y), \lambda x_1 + \mu x_2 \rangle = -\lambda \langle f(y), x_1 \rangle - \mu \langle f(y), x_2 \rangle$$
$$= \lambda \langle f(x_1), y \rangle + \mu \langle f(x_2), y \rangle,$$

which implies

$$\langle f(\lambda x_1 + \mu x_2) - \lambda f(x_1) - \mu f(x_2), y \rangle = 0$$

for any x_1, x_2 and y in H and any λ, μ in $\mathbb{R}$, wherefrom we have the linearity of f. Because $\langle f(x) - f(y), x - y \rangle = 0$, for any x, y in H, f is also skew-adjoint. Thus $F(x) = f(x) + F(0)$ and hence it is indeed affine with skew-adjoint linear term.

The implication $2 \Rightarrow 1$ is obvious $\qquad\qquad\qquad\qquad\qquad\qquad\square$

3.1.3   Scalar Derivatives and Scalar Differentiability

DEFINITION 3.5 *Let $p > 0$ and $C_1, C_2 \subseteq H$ such that 0 is a nonisolated point of C_1 and x_0 a nonisolated point of C_2. Let $f, g : C_2 \to H$. The following definition is an extension of Definition 1.6. The limit*

$$\underline{(f, g)}_p^{\#}(x_0, C_1) = \liminf_{\substack{x \to x_0 \\ x - x_0 \in C_1}} \frac{\langle f(x) - f(x_0), g(x) - g(x_0) \rangle}{\|x - x_0\|^p}$$

is called the lower scalar derivative of rank p *of the (unordered) pair of mappings (f, g) in x_0 in the direction of C_1. Taking* lim sup *in place of* lim inf*, we can define the* upper scalar derivative of rank p *of the (unordered) pair of mappings (f, g) at x_0 in the direction of C_1 similarly. It is denoted $\overline{(f, g)}_p^{\#}(x_0, C_1)$. If $C_1 = H$ or $C_1 = C_2$ and $x_0 = 0$, then without confusion, we can omit the phrase "in the direction of C_1" from the definitions. In this case, we omit C_1*

from the corresponding notations. If $p = 2$, then we omit the phrase "of rank p" from the definitions and p from the subscript of the corresponding notations. If g is the inclusion mapping of C_2 into H we speak in short about the scalar derivatives of the mapping f and drop g from the corresponding notations.

REMARK 3.1 *If x_0 is an interior point of C_2, $C_2 \subseteq \tilde{C}_2$ and $F, G : \tilde{C}_2 \to H$ such that $f = F|_{C_2}$ and $g = G|_{C_2}$, then $\underline{(f, g)}_p^{\#}(x_0, C_1) = \underline{(F, G)}_p^{\#}(x_0, C_1)$.*

The next remark follows directly from the definitions of the lower and upper scalar derivatives of rank p along a set.

REMARK 3.2 *Let $p > 0$, $x \in H$, and $C \subseteq H$ such that 0 is a nonisolated point of C. Let $\Psi, \Psi_1, \Psi_2 : H \to H$ be additive mappings. Then, $\underline{(\Psi_1, \Psi_2)}_p^{\#}(\cdot, C) : H \to \overline{\mathbb{R}}$ and $\overline{(\Psi_1, \Psi_2)}_p^{\#}(\cdot, C) : H \to \overline{\mathbb{R}}$ are constant functions where $\overline{\mathbb{R}} = \mathbb{R} \cup \{-\infty, \infty\}$. In particular, we have that $\underline{\Psi}_p^{\#}(\cdot, C) : H \to \overline{\mathbb{R}}$ and $\overline{\Psi}_p^{\#}(\cdot, C) : H \to \overline{\mathbb{R}}$ are constant functions.*

DEFINITION 3.6 *Consider the operator $f : H \to H$. If the limit*

$$\lim_{x \to x_0} \frac{\langle f(x) - f(x_0), x - x_0 \rangle}{\|x - x_0\|^2} =: f^{\#}(x_0) \in \mathbb{R}$$

exists (here $\|x - x_0\|^2 = \langle x - x_0, x - x_0 \rangle$), then it is called the scalar derivative *of the operator f in x_0. In this case f is said to be* scalarly differentiable *at x_0. If $f^{\#}(x)$ exists for every x in H, then f is said to be* scalarly differentiable *on H, with the* scalar derivative $f^{\#}$.

It follows from this definition that both the set of operators scalarly differentiable in x_0, and the set of operators scalarly differentiable on H form linear spaces.

THEOREM 3.7 *The linear operator $A : H \to H$ is scalarly differentiable on H if and only if it is of the form $A = B + cI$ with B an skew-adjoint linear operator, I the identity of H, and c a real number.*

Proof. Let us suppose that A is scalarly differentiable in $x_0 \in H$. Then

$$A^{\#}(x_0) = \liminf_{x \to x_0} \frac{\langle Ax - Ax_0, x - x_0 \rangle}{\|x - x_0\|^2} = \liminf_{v \to 0} \frac{\langle Av, v \rangle}{\|v\|^2} = A^{\#}(0).$$

Take $v = \lambda x$ with $x \in H$ and $\lambda > 0$. Then

$$A^{\#}(0) = \liminf_{\lambda \downarrow 0} \frac{\langle A\lambda x, \lambda x \rangle}{\|\lambda x\|^2} = \frac{\langle Ax, x \rangle}{\|x\|^2}.$$

That is, $\langle Ax, x \rangle / \|x\|^2 = c = A^{\#}(0)$. Accordingly,

$$\langle (A - cI)x, x \rangle = 0, \ \forall x \in H.$$

This means that $B = A - cI$ is a skew-adjoint linear operator and hence A has the representation given in the theorem. Obviously, every $A = B + cI$ with B a skew-adjoint linear operator has the scalar derivative c at every point of H. $\square$

THEOREM 3.8 *Suppose that* $f : H \to H$ *is both Gateaux differentiable and scalarly differentiable in* x_0. *Then we have for the Gateaux differential* $\delta f(x_0)$ *of* f *at* x_0 *the relation*

$$\delta f(x_0) = B + f^{\#}(x_0)I,$$

with $B : H \to H$ *linear and skew-adjoint.*

Proof. Let $t \in H$ be given. Then

$$f^{\#}(x_0) = \frac{1}{\|t\|^2} \liminf_{\lambda \downarrow 0} \left\langle \frac{f(x_0 + \lambda t) - f(x_0)}{\lambda}, t \right\rangle = \frac{1}{\|t\|^2} \langle \delta f(x_0)(t), t \rangle,$$

wherefrom $\langle (\delta f(x_0) - f^{\#}(x_0)I)(t), t \rangle = 0, \ \forall t \in H$; that is,

$$B = \delta f(x_0) - f^{\#}(x_0)I$$

is linear and skew-adjoint. $\square$

3.1.4 Characterization of Monotone Mappings by Using Scalar Derivatives

By using the notion of the upper (lower) scalar derivative we obtain the following assertion.

THEOREM 3.9 *Let* C *be an open convex set in* H, $p > 2$, *and* $f : C \to H$. *Then the following statements are equivalent.*

1. f $(-f)$ *is a monotone operator.*

2. $\underline{f}_p^{\#}(x) \geq 0$ $(\overline{f}_p^{\#}(x) \leq 0)$ *for each* x *in* C.

Proof. The implication $1 \Rightarrow 2$ is obvious.
$2 \Rightarrow 1$ Take $\varepsilon > 0$ arbitrarily and put $g = f + \varepsilon I$. Then

$$\underline{g}_p^{\#}(x) \geq \underline{f}_p^{\#}(x) + \varepsilon \liminf_{y \to x} \|y - x\|^{2-p} > 0 \ \ \forall x \in C.$$

Take a, b in C, $a \neq b$. For x in the line segment $[a, b]$ determined by a and b, one has by hypothesis:

$$\liminf_{y \to x} \frac{\langle g(y) - g(x), y - x \rangle}{\|y - x\|^p} > 0,$$

and hence there exists $\delta(x) > 0$ such that for any y in

$$I_x =]x - \delta(x)(b-a), x + \delta(x)(b-a)[\subset C,$$

$\langle g(y) - g(x), y - x \rangle > 0$ holds as far as $y \neq x$. Obviously,

$$[a, b] \subset \bigcup_{x \in [a,b]} I_x;$$

that is, $\{I_x : x \in [a, b]\}$ is an open cover of the compact set $[a, b]$. Hence,

$$[a, b] \subset I_{y_1} \cup I_{y_2} \cup \cdots \cup I_{y_{m-1}}$$

for an appropriate set $y_1, \ldots, y_{m-1}$ of points in $[a, b]$. We can suppose that $y_1, \ldots, y_{m-1}$ are ordered from a to b. Hence, $a = y_0 \in I_{y_1}$, $b = y_m \in I_{y_{m-1}}$. We can also consider that no interval I_{y_i} is contained in any other. Take $\xi_i \in I_{y_{i-1}} \cap I_{y_i} \cap]y_{i-1}, y_i[$. Then, by the construction of these intervals

$$\langle g(\xi_i) - g(y_{i-1}), \xi_i - y_{i-1} \rangle > 0,$$

$$\langle g(y_i) - g(\xi_i), y_i - \xi_i \rangle > 0$$

and because ξ_i is in $]y_{i-1}, y_i[$,

$$y_i - \xi_i = \alpha(y_i - y_{i-1}),$$

$$\xi_i - y_{i-1} = \beta(y_i - y_{i-1}),$$

for appropriate positive α and β. Hence,

$$\langle g(\xi_i) - g(y_{i-1}), y_i - y_{i-1} \rangle > 0,$$

$$\langle g(y_i) - g(\xi_i), y_i - y_{i-1} \rangle > 0,$$

wherefrom

$$\langle g(y_i) - g(y_{i-1}), y_i - y_{i-1} \rangle > 0.$$

But $y_i - y_{i-1} = \lambda_i(b - a)$ for some positive λ_i, and then we must also have

$$\langle g(y_i) - g(y_{i-1}), b - a \rangle > 0.$$

By summing the above relations from $i = 1$ to $i = m$, we obtain

$$\langle g(b) - g(a), b - a \rangle > 0.$$

Rewriting this relation by using the definition of g we have

$$\langle f(b) - f(a), b - a \rangle + \varepsilon \|b - a\|^2 > 0.$$

By letting $\varepsilon \to 0$ we conclude that

$$\langle f(b) - f(a), b - a \rangle \geq 0.$$

The case $\overline{f}_p^{\#}(x) \leq 0, \forall x \in C$ can be handled similarly. $\square$

THEOREM 3.10 *Let C be an open convex set in H, $p > 0$ and suppose that $f : C \to H$ satisfies*

$$\underline{f}_p^{\#}(x) > 0 \ (\overline{f}_p^{\#}(x) < 0), \ \forall x \in C.$$

Then $f \ (-f)$ is strictly monotone on C.

This theorem can be proved by an appropriate adaptation of the proof of Theorem 3.9. A careful analysis shows that in this case we only need $p > 0$.

EXAMPLE 3.11 *A classical counterexample in calculus is the function $f : \mathbb{R} \to \mathbb{R}$; $f(x) = x^3$ which is strictly monotone, but $\underline{f}^{\#}(0) = f'(0) = 0$. This is presented as an example of a function which is strictly monotone and whose derivative is not positive everywhere. However, for $p = 4$ we have $\underline{f}_4^{\#}(x) = \infty$ for every $x \in \mathbb{R}\backslash\{0\}$ and $\underline{f}_4^{\#}(0) = 1$. Therefore, $\underline{f}_4^{\#}(x) > 0$ for every $x \in \mathbb{R}$. Obviously, this remark can be extended to x^{2k+1}, for any positive integer k.*

COROLLARY 3.12 *Let $f : H \to H$ be given. The following statements are equivalent.*

1. *f possesses bounded (upper and lower) scalar p-derivatives with $p > 2$ at each point of H.*

2. *$f^{\#}(x) = 0, \forall x \in H$.*

3. *f is an affine function with skew-adjoint linear part.*

Proof. The implication $3 \Rightarrow 1$ is trivial.

Next we prove $1 \Rightarrow 2$ Suppose that $\max\{\|\underline{f}_p^{\#}(x)\|, \|\overline{f}_p^{\#}(x)\|\} < \rho_x$, for some $\rho_x \in \mathbb{R}_+$. Then, for $\delta > 0$ there exists $\varepsilon > 0$ such that

$$\left| \frac{\langle f(y) - f(x), y - x \rangle}{\|y - x\|^2} \right| < (\rho_x + \delta)\|y - x\|^{p-2},$$

whenever $\|y - x\| < \varepsilon$. Passing with y to x we get that $f^{\#}(x)$ exists and is zero.

To show that $2 \Rightarrow 3$ we apply Theorem 3.9 to conclude that

$$\langle f(y) - f(x), y - x \rangle = 0,$$

$\forall x, y \in H$, and then by usage of Theorem 3.4 we conclude assertion 3. $\square$

3.1.5   Computational Formulae for the Scalar Derivatives

We recall that a subset of a Hilbert space is called a cone if it is invariant under multiplication by positive scalars, and a cone is called a convex cone if it is invariant under addition. The first theorem of this section shows that the lower and upper scalar derivatives of a pair of Frechét differentiable mappings in a point x are equal to the lower and upper scalar derivatives in 0 of the pair of their differentials in x, respectively.

THEOREM 3.13 *Let* $x \in H$ *and* $K \subseteq H$ *be a closed cone. If* $f, g : H \to H$ *are Frechét differentiable in* x, *with the differentials* $df(x), dg(x)$, *respectively, then*

$$\underline{(f,g)^{\#}}(x, K) = \underline{(df(x), dg(x))^{\#}}(0, K),$$

$$\overline{(f,g)^{\#}}(x, K) = \overline{(df(x), dg(x))^{\#}}(0, K).$$

Proof. For the expressions

$$\frac{f(x+v) - f(x) - df(x)(v)}{\|v\|} = \omega(f)(x, v)$$

and

$$\frac{g(x+v) - g(x) - dg(x)(v)}{\|v\|} = \omega(g)(x, v)$$

we have $\lim_{v \to 0} \omega(f)(x, v) = 0$ and $\lim_{v \to 0} \omega(g)(x, v) = 0$, hence

$$\underline{(f,g)^{\#}}(x, K) = \liminf_{\substack{v \to 0 \\ v \in K}} \frac{\langle f(x+v) - f(x), g(x+v) - g(x) \rangle}{\|v\|^2}$$

$$= \liminf_{\substack{v \to 0 \\ v \in K}} \left\langle \omega(f)(x, v) + df(x)\left(\frac{v}{\|v\|}\right), \omega(g)(x, v) + dg(x)\left(\frac{v}{\|v\|}\right) \right\rangle$$

$$= \lim_{\substack{v \to 0 \\ v \in K}} \langle \omega(f)(x, v), \omega(g)(x, v) \rangle + \lim_{\substack{v \to 0 \\ v \in K}} \left\langle \omega(f)(x, v), dg(x)\left(\frac{v}{\|v\|}\right) \right\rangle$$

$$+ \lim_{\substack{v \to 0 \\ v \in K}} \left\langle df(x)\left(\frac{v}{\|v\|}\right), \omega(g)(x, v) \right\rangle$$

$$+ \liminf_{\substack{v \to 0 \\ v \in K}} \left\langle df(x)\left(\frac{v}{\|v\|}\right), dg(x)\left(\frac{v}{\|v\|}\right) \right\rangle.$$

We used in the last relation the fact that the first three limits exist because

$$|\langle \omega(f)(x, v), \omega(g)(x, v) \rangle| \leq \|\omega(f)(x, v)\| \cdot \|\omega(g)(x, v)\| \to 0$$

for $v \to 0$, $v \in K$,

$$\left| \left\langle \omega(f)(x,v), dg(x)\left(\frac{v}{\|v\|}\right) \right\rangle \right| \leq \|\omega(f)(x,v)\| \cdot \|d(g)(x)\| \to 0$$

for $v \to 0$, $v \in K$, and

$$\left| \left\langle df(x)\left(\frac{v}{\|v\|}\right), \omega(g)(x,v) \right\rangle \right| \leq \|d(f)(x)\| \cdot \|\omega(g)(x,v)\| \to 0$$

for $v \to 0$, $v \in K$. Hence,

$$\underline{(f,g)^{\#}}(x) = \liminf_{\substack{v \to 0 \\ v \in K}} \frac{\langle df(x)(v), dg(x)(v)\rangle}{\|v\|^2} = \underline{(df(x), dg(x))^{\#}}(0, K).$$

A similar way yields the proof of the second relation of the theorem. □

In particular, we have as follows.

THEOREM 3.14 *Let $x \in H$ and $K \subseteq H$ be a closed cone. If $f : H \to H$ is Frechét differentiable in x, with the differential $df(x)$, then*

$$\underline{f^{\#}}(x, K) = \underline{df(x)^{\#}}(0, K),$$

$$\overline{f}^{\#}(x, K) = \overline{df(x)}^{\#}(0, K).$$

The next theorem can be proved similarly to Theorem 3.13.

THEOREM 3.15 *Let $K \subseteq H$ be a closed convex cone with nonempty interior and x an interior point of K. If $f, g : K \to H$ are Frechét differentiable in x, with differentials $df(x)$ and $dg(x)$, respectively, then*

$$\underline{(f,g)^{\#}}(x, K) = \underline{(df(x), dg(x))^{\#}}(0, K),$$

$$\overline{(f,g)}^{\#}(x, K) = \overline{(df(x), dg(x))}^{\#}(0, K).$$

In particular, we have as follows.

THEOREM 3.16 *Let $K \subseteq H$ be a closed convex cone with nonempty interior and x an interior point of K. If $f : K \to H$ is Frechét differentiable in x, with the differential $df(x)$, then*

$$\underline{f^{\#}}(x, K) = \underline{df(x)^{\#}}(0, K),$$

$$\overline{f}^{\#}(x, K) = \overline{df(x)}^{\#}(0, K).$$

The next theorem gives the computational formulae for the lower and upper scalar derivatives of a pair of positively homogeneous mappings in 0.

THEOREM 3.17 *Let $K \subseteq H$ be a closed cone. If $\Phi_1, \Phi_2 : H \to H$ are positively homogeneous, then*

$$\underline{(\Phi_1, \Phi_2)}^{\#}(0, K) = \inf_{\substack{\|u\|=1 \\ u \in K}} \langle \Phi_1(u), \Phi_2(u) \rangle,$$

$$\overline{(\Phi_1, \Phi_2)}^{\#}(0, K) = \sup_{\substack{\|u\|=1 \\ u \in K}} \langle \Phi_1(u), \Phi_2(u) \rangle.$$

Proof. We prove only the first equality. The second equality can be proved similarly. We have

$$\underline{(\Phi_1, \Phi_2)}^{\#}(0, K) = \liminf_{\substack{v \to 0 \\ v \in K}} \frac{\langle \Phi_1(v), \Phi_2(v) \rangle}{\|v\|^2} \leq \liminf_{t \downarrow 0} \frac{\langle \Phi_1(tu), \Phi_2(tu) \rangle}{\|tu\|^2}$$

$$= \langle \Phi_1(u), \Phi_2(u) \rangle,$$

for all $u \in K$ with $\|u\| = 1$. Therefore,

$$\underline{(\Phi_1, \Phi_2)}^{\#}(0, K) \leq \inf_{\substack{\|u\|=1 \\ u \in K}} \langle \Phi_1(u), \Phi_2(u) \rangle.$$

Conversely,

$$\frac{\langle \Phi_1(v), \Phi_2(v) \rangle}{\|v\|^2} = \left\langle \Phi_1\left(\frac{v}{\|v\|}\right), \Phi_2\left(\frac{v}{\|v\|}\right) \right\rangle \geq \inf_{\substack{\|u\|=1 \\ u \in K}} \langle \Phi_1(u), \Phi_2(u) \rangle,$$

for all $v \in K \backslash \{0\}$, which implies

$$\underline{(\Phi_1, \Phi_2)}^{\#}(0, K) = \liminf_{\substack{v \to 0 \\ v \in K}} \frac{\langle \Phi_1(v), \Phi_2(v) \rangle}{\|v\|^2} \geq \inf_{\substack{\|u\|=1 \\ u \in K}} \langle \Phi_1(u), \Phi_2(u) \rangle.$$

Hence,

$$\underline{(\Phi_1, \Phi_2)}^{\#}(0, K) = \inf_{\substack{\|u\|=1 \\ u \in K}} \langle \Phi_1(u), \Phi_2(u) \rangle.$$

$\square$

In particular, we have the following.

THEOREM 3.18 *Let $K \subseteq H$ be a closed cone. If $\Phi : H \to H$ is positively homogeneous, then*

$$\underline{\Phi}^{\#}(0, K) = \inf_{\substack{\|u\|=1 \\ u \in K}} \langle \Phi(u), u \rangle,$$

$$\overline{\Phi}^{\#}(0, K) = \sup_{\substack{\|u\|=1 \\ u \in K}} \langle \Phi(u), u \rangle.$$

The next theorem can be proved similarly to Theorem 3.17.

THEOREM 3.19 *Let $K \subseteq H$ be a closed convex cone. If $\Phi_1, \Phi_2 : K \to H$ are positively homogeneous, then*

$$\underline{(\Phi_1, \Phi_2)}^{\#}(0) = \inf_{\substack{\|u\|=1 \\ u \in K}} \langle \Phi_1(u), \Phi_2(u) \rangle,$$

$$\overline{(\Phi_1, \Phi_2)}^{\#}(0) = \sup_{\substack{\|u\|=1 \\ u \in K}} \langle \Phi_1(u), \Phi_2(u) \rangle.$$

In particular, we have as follows.

THEOREM 3.20 *Let $K \subseteq H$ be a closed convex cone. If $\Phi : K \to H$ is positively homogeneous, then*

$$\underline{\Phi}^{\#}(0) = \inf_{\substack{\|u\|=1 \\ u \in K}} \langle \Phi(u), u \rangle,$$

$$\overline{\Phi}^{\#}(0) = \sup_{\substack{\|u\|=1 \\ u \in K}} \langle \Phi(u), u \rangle.$$

The remaining results of the section give effective computational formulae which can be used to verify those conditions in [Isac and Nemeth, 2003, 2004, 2005a,b, 2006a,b, 2007a,b] which are expressed with the scalar derivative.

Theorems 3.13 and 3.17 imply the following.

THEOREM 3.21 *Let $x \in H$ and $K \subseteq H$ be a closed cone. If $f, g : H \to H$ are Frechét differentiable in x, with differentials $df(x)$ and $dg(x)$, respectively, then*

$$\underline{(f, g)}^{\#}(x, K) = \inf_{\substack{\|u\|=1 \\ u \in K}} \langle df(x)(u), dg(x)(u) \rangle,$$

$$\overline{(f, g)}^{\#}(x, K) = \sup_{\substack{\|u\|=1 \\ u \in K}} \langle df(x)(u), dg(x)(u) \rangle.$$

Theorems 3.15 and 3.19 imply the following.

THEOREM 3.22 *Let $K \subseteq H$ be a closed convex cone with nonempty interior and x an interior point of K. If $f, g : K \to H$ are Frechét differentiable in x, with differentials $df(x)$ and $dg(x)$, then*

$$\underline{(f,g)}^{\#}(x, K) = \inf_{\substack{\|u\|=1 \\ u \in K}} \langle df(x)(u), dg(x)(u) \rangle,$$

$$\overline{(f,g)}^{\#}(x, K) = \sup_{\substack{\|u\|=1 \\ u \in K}} \langle df(x)(u), dg(x)(u) \rangle.$$

Theorems 3.14 and 3.18 imply the following.

THEOREM 3.23 *Let $x \in H$ and $K \subseteq H$ be a closed cone. If $f : H \to H$ is Frechét differentiable in x, with the differential $df(x)$, then*

$$\underline{f}^{\#}(x, K) = \inf_{\substack{\|u\|=1 \\ u \in K}} \langle df(x)(u), u \rangle,$$

$$\overline{f}^{\#}(x, K) = \sup_{\substack{\|u\|=1 \\ u \in K}} \langle df(x)(u), u \rangle.$$

Theorems 3.16 and 3.20 imply the following.

THEOREM 3.24 *Let $K \subseteq H$ be a closed convex cone with nonempty interior and x an interior point of K. If $f : K \to H$ is Frechét differentiable in x, with the differential $df(x)$, then*

$$\underline{f}^{\#}(x, K) = \inf_{\substack{\|u\|=1 \\ u \in K}} \langle df(x)(u), u \rangle,$$

$$\overline{f}^{\#}(x, K) = \sup_{\substack{\|u\|=1 \\ u \in K}} \langle df(x)(u), u \rangle.$$

3.2   Inversions

Let $(H, \langle \cdot, \cdot \rangle)$ be a Hilbert space and $\| \cdot \|$ the norm generated by $\langle \cdot, \cdot \rangle$. The following definition is an extension of Example 5.1, p.169 of [do Carmo, 1992]:

DEFINITION 3.25 *The operator*

$$i : H \backslash \{0\} \to H \backslash \{0\}; \quad i(x) = \frac{x}{\|x\|^2}$$

is called the inversion *(of pole 0).*

It is easy to see that i is one to one and $i^{-1} = i$. Indeed, because $\|i(x)\| = 1/\|x\|$, by the definition of i we have $i(i(x)) = (i(x))/\|i(x)\|^2 = \|x\|^2 i(x) = x$.

Hence i is a global diffeomorphism of $H\backslash\{0\}$ which can be viewed as a global nonlinear coordinate transformation in H.

Let $A \subseteq H$ such that $0 \in A$ and $A\backslash\{0\}$ is an invariant set of the inversion i; that is, $i(A\backslash\{0\}) = A\backslash\{0\}$ and $f : A \to H$. Examples of invariant sets of the inversion i are:

1. $F\backslash\{0\}$ where F is a linear subspace of H (in particular F can be the whole H),

2. $K\backslash\{0\}$ where $K \subseteq H$ is a pointed convex cone.

Now we define the inversion (of pole 0) of the mapping f.

DEFINITION 3.26 *The* inversion *(of pole 0) of the mapping f is the mapping* $\mathcal{I}(f) : A \to H$ *defined by:*

$$\mathcal{I}(f)(x) = \begin{cases} \|x\|^2(f \circ i)(x) & \text{if} \quad x \neq 0, \\ 0 & \text{if} \quad x = 0. \end{cases}$$

PROPOSITION 3.1 *The inversion of mappings $\mathcal{I}$ is a one-to-one operator on the set of mappings $\{f| f : A \to H; f(0) = 0\}$ and $\mathcal{I}^{-1} = \mathcal{I}$; that is,* $\mathcal{I}(\mathcal{I}(f)) = f$.

Proof. By definition $\mathcal{I}(\mathcal{I}(f))(0) = 0$. Hence, $\mathcal{I}(\mathcal{I}(f))(0) = f(0)$. If $x \neq 0$ then $\mathcal{I}(\mathcal{I}(f))(x) = \|x\|^2\mathcal{I}(f)(i(x)) = \|x\|^2\|i(x)\|^2 f(i(i(x))) = f(x)$. Thus, $\mathcal{I}(\mathcal{I}(f))(x) = f(x)$ for all $x \in K$. Therefore $\mathcal{I}(\mathcal{I}(f)) = f$. $\square$

PROPOSITION 3.2 *Let $f : A \to A$. Then, $x \neq 0$ is a fixed point of f iff $i(x)$ is a fixed point of $\mathcal{I}(f)$.*

Proof. Suppose that $x \neq 0$ is a fixed point of f; that is, $f(x) = x$. Because $i(i(x)) = x$ we have

$$f(i(i(x))) = x. \tag{3.1}$$

Multiplying (3.1) by $\|i(x)\|^2 = 1/\|x\|^2$ we obtain $\mathcal{I}(f)(i(x)) = i(x)$. Thus, $i(x)$ is a fixed point of $\mathcal{I}(f)$. Similarly, it can be proved that if $i(x)$ is a fixed point of $\mathcal{I}(f)$, then x is a fixed point of f. $\square$

Let $D = \{x \in H \mid \|x\| \leq 1\}$ and $C = \{x \in H \mid \|x\| = 1\}$ be the unit ball and the unit sphere of H, respectively.

PROPOSITION 3.3 *Let $f, g : A \to H$ such that $f(x) = g(x)$, for all $x \in A \cap C$ and $f(0) = g(0) = 0$. There exist unique extensions $\tilde{f}, \tilde{g} : A \to H$ of $f|_{A \cap D}$ and $g|_{A \cap D}$, respectively, such that $\tilde{g} = \mathcal{I}(\tilde{f})$.*

Proof. Let $D^\circ = \{x \in H \mid \|x\| < 1\}$. First we prove the existence of the extensions $\tilde{f}, \tilde{g}$. Define the extensions $\tilde{f}, \tilde{g}$ of $f|_{A \cap D}$ and $g|_{A \cap D}$ by

$$\tilde{g}(x) = \begin{cases} g(x) & \text{if } \|x\| \leq 1 \\ \mathcal{I}(f)(x) & \text{if } \|x\| > 1 \end{cases}$$

and

$$\tilde{f}(x) = \begin{cases} f(x) & \text{if } \|x\| \leq 1 \\ \mathcal{I}(g)(x) & \text{if } \|x\| > 1 \end{cases},$$

respectively. We have to prove that

$$\tilde{g}(x) = \mathcal{I}(\tilde{f})(x) \tag{3.2}$$

for all $x \in A$. We consider three cases.

First case: $x \in A \cap D^\circ$. In this case $\|x\| < 1$ and hence, $\|i(x)\| > 1$. Thus, by definition $\tilde{g}(x) = g(x)$ and $\tilde{f}(i(x)) = \mathcal{I}(g)(i(x))$. By using these relations and the definition of the inversion of a mapping, relation (3.2) can be proved easily.

Second case: $x \in A \setminus D$. In this case $\|x\| > 1$ and hence, $\|i(x)\| < 1$. Thus, by definition $\tilde{g}(x) = \mathcal{I}(f)(x)$ and $\tilde{f}(i(x)) = f(i(x))$. Relation (3.2) can be proved similarly to the previous case.

Third case: $x \in A \cap C$. In this case $\|x\| = 1$ and hence, $i(x) = x$. Thus, by definition $\tilde{g}(x) = g(x)$ and $\tilde{f}(i(x)) = f(x)$. In this case (3.2) is equivalent to $f(x) = g(x)$, which by the assumption made on f and g it is true.

Now we prove the uniqueness of the extensions $\tilde{f}, \tilde{g}$. Suppose that $\hat{f}, \hat{g}$ are extensions of $f|_{A \cap D}$ and $g|_{A \cap D}$, respectively, such that $\hat{g} = \mathcal{I}(\hat{f})$. If $\|x\| \leq 1$, then $\hat{g}(x) = \tilde{g}(x) = g(x)$ because both $\hat{g}$ and $\tilde{g}$ are extensions of $g|_{A \cap D}$. If $\|x\| > 1$, then $\|i(x)\| < 1$. Because $\hat{f}$ is an extension of $f|_{A \cap D}$, $\hat{f}(i(x)) = f(i(x))$. By using this relation, relation $\hat{g}(x) = \mathcal{I}(\hat{f})(x)$, the definition of the inversion of a mapping, and the definition of $\tilde{g}$, we obtain $\hat{g}(x) = \tilde{g}(x)$. Hence, $\hat{g} = \tilde{g}$. Relation $\hat{g} = \mathcal{I}(\hat{f})$ implies $\hat{f} = \mathcal{I}(\hat{g})$. Hence relation $\hat{f} = \tilde{f}$ can be proved by interchanging the roles of f and g. $\qquad \square$

In the case of $f = g$ Proposition 3.3 has the following corollary.

COROLLARY 3.27 *Let $f : A \to H$; $f(0) = 0$. There exists a unique extension $\tilde{f} : A \to H$ of $f|_{A \cap D}$ such that $\tilde{f}$ is a fixed point of $\mathcal{I}$ (i.e., $\tilde{f} = \mathcal{I}(\tilde{f})$).*

It is easy to see that the inversion of mappings is linear, that if $T \in L(H, H)$ and $j : A \hookrightarrow H$ is the embedding of A into H then $\mathcal{I}(T \circ j) = T \circ j$ and that if $\|x\| \to +\infty$ then $i(x) \to 0$.

We have as follows.

LEMMA 3.28 *Let $K \subseteq H$ be an unbounded set such that $0 \in K$ and $K \backslash \{0\}$ is an invariant set of the inversion i. Let $g : H \to H$. Then we have*

$$\liminf_{\substack{\|x\| \to \infty \\ x \in K}} \frac{\langle g(x), x \rangle}{\|x\|^2} = \underline{\mathcal{I}(g)}^{\#,K}(0).$$

Proof. Because $K \subseteq H$ is unbounded and $K \backslash \{0\}$ is an invariant set of i, 0 is a nonisolated point of K. Hence, $\underline{\mathcal{I}(g)}^{\#,K}(0)$ is well defined. Consider the global nonlinear coordinate transformation $y = i(x)$. Then, $x = i(y)$ and we have

$$\liminf_{\substack{\|x\| \to \infty \\ x \in K}} \frac{\langle g(x), x \rangle}{\|x\|^2} = \liminf_{\substack{y \to 0 \\ y \in K}} \langle \mathcal{I}(g)(y), i(y) \rangle,$$

from where, by using the definition of the lower scalar derivative along a set, the assertion of the lemma follows easily.

3.3 Fixed Point Theorems Generated by Krasnoselskii's Fixed Point Theorem

Let $(H, \langle \cdot, \cdot \rangle)$ be a Hilbert space, $K \subseteq H$ a generating closed pointed convex cone and $f : K \to K$. If in Theorem 2.38 we replace assumptions 1. and 2. by "1. f is completely continuous", we obtain the following.

THEOREM 3.29 *If the following assumptions are satisfied.*

1. *f is completely continuous;*

2. *There exists an asymptotically scalarly differentiable mapping $f_0 : K \to H$ such that $f_0 : K \to H$, $f \leq_{K^*} f_0$, and $\|f'_{0,s}(\infty)\| < 1$;*

then f has a fixed point.

By Theorem 3.29 and Theorem 2.33 we have the following fixed point theorem.

THEOREM 3.30 *If the following assumptions are satisfied,*

1. *f is completely continuous;*

2. *There exists a mapping $f_0 : K \to H$ such that $f \leq_{K^*} f_0$ and $\overline{\mathcal{I}(f_0)}^{\#}(0) < 1$;*

then f has a fixed point.

Proof. By Theorem 2.33 the linear operator $T = \overline{\mathcal{I}(f_0)}^{\#}(0)I$ is an asymptotic scalar derivative of f_0. We have $\|T\| = |\overline{\mathcal{I}(f_0)}^{\#}(0)|$. We consider two cases.

1. $\overline{\mathcal{I}(f_0)}^{\#}(0) \leq 0$. In this case choose a $c \in]-1, 0] \cap [\overline{\mathcal{I}(f_0)}^{\#}(0), +\infty[$. By Remark 2.15, $T = cI$ is an asymptotic scalar derivative of f_0 with $\|T\| = -c < 1$.

2. $0 < \overline{\mathcal{I}(f_0)}^{\#}(0) < 1$. In this case $\|T\| = \overline{\mathcal{I}(f_0)}^{\#}(0) < 1$.

It follows that $\|T\| < 1$. By using Theorem 3.29, f has a fixed point. $\square$

COROLLARY 3.31 *If the following assumptions are satisfied,*

1. *f is completely continuous;*

2. *$\overline{\mathcal{I}(f)}^{\#}(0) < 1$;*

then f has a fixed point.

Corollary 3.31 has the following interesting consequence.

PROPOSITION 3.4 *Let $q : K \to K$ be a completely continuous mapping such that $I \leq_K q$ and $f : K \to K$; $f = q - I$. Then, $\overline{\mathcal{I}(f)}^{\#}(0) \geq 0$.*

Proof. Suppose that $\overline{\mathcal{I}(f)}^{\#}(0) < 0$. Because K is generating $K \neq \{0\}$. Let $a \in K\backslash\{0\}$. Because $K + K \subseteq K$, $x + f(x) + a \in K$ for all $x \in K$. Define $q_a : K \to K$ by $q_a(x) = x + f(x) + a$. Because $q_a = q + a$, q_a is completely continuous. We also have $\overline{\mathcal{I}(q_a)}^{\#}(0) = 1 + \overline{\mathcal{I}(f)}^{\#}(0) < 1$. Hence, by Corollary 3.31, q_a has a fixed point; that is, the equation $f(x) = -a$ has a solution. It follows that $a \in -K$. Because $K \cap (-K) = \{0\}$, it follows that $a = 0$. But this is in contradiction with $a \in K\backslash\{0\}$. Hence, $\overline{\mathcal{I}(f)}^{\#}(0) \geq 0$. $\square$

3.4   Surjectivity Theorems

Let $(H, \langle\cdot, \cdot\rangle)$ be a Hilbert space, $K \subseteq H$ a generating closed pointed convex cone, and $f : K \to K$.

THEOREM 3.32 *If the following assumptions are satisfied,*

1. *$f = I - q$, where $q : K \to K$ is completely continuous and $q \leq_K I$;*

2. *There exists a mapping $f_0 : K \to H$ such that $f_0 \leq_{K*} f$ and $\underline{\mathcal{I}(f_0)}^{\#}(0) > 0$;*

then f is surjective.

Proof. Let $y \in K$ be arbitrary but fixed. Define the mapping $q_{y,0} : K \to H$ by $q_{y,0} = x - f_0(x) + y$. Because $K + K \subseteq K$, $x - f(x) + y = q(x) + y \in K$ for all $x \in K$. Define the mapping $q_y : K \to K$ by $q_y(x) = x - f(x) + y$. It is easy to see that q_y is completely continuous, $q_y \leq_{K^*} q_{y,0}$, and

$$\overline{\mathcal{I}(q_{y,0})}^{\#}(0) = 1 - \underline{\mathcal{I}(f_0)}^{\#}(0) < 1.$$

Hence, by Theorem 3.30, q_y has a fixed point; that is, the equation $f(x) = y$ has a solution. Because y was arbitrarily chosen, f is surjective. $\square$

COROLLARY 3.33 *If the following assumptions are satisfied,*

1. $f = I - q$, *where* $q : K \to K$ *is completely continuous and* $q \leq_K I$;

2. $\underline{\mathcal{I}(f)}^{\#}(0) > 0$;

then f is surjective.

THEOREM 3.34 *If the following assumptions are satisfied,*

1. $f = bI - q$, *where* $b > 0$, $q : K \to K$ *is completely continuous and* $q \leq_K bI$;

2. *There exists a mapping* $f_0 : K \to H$ *such that* $f_0 \leq_{K^*} f$ *and* $\underline{\mathcal{I}(f_0)}^{\#}(0) > 0$;

then f is surjective.

Proof. By using Theorem 3.32 with $(1/b)f_0$, $(1/b)f$, and $(1/b)q$ replacing f_0, f, and q, respectively, we obtain that $(1/b)f$ is surjective. Hence, f is also surjective. $\square$

COROLLARY 3.35 *If the following assumptions are satisfied,*

1. $f = bI - q$, *where* $b > 0$, $q : K \to K$ *is completely continuous and* $q \leq_K bI$;

2. $\underline{\mathcal{I}(f)}^{\#}(0) > 0$;

then f is surjective.

LEMMA 3.36 *Let* $A \subseteq H$ *such that* $A \setminus \{0\}$ *is an invariant set of the inversion i and* $\Upsilon = \{\tau \mid \tau : A \to H\}$. *The inversion of mappings* $\mathcal{I}$ *is* K^*-*monotone on* Υ; *that is,* $\mathcal{I}(\tau_1) \leq_{K^*} \mathcal{I}(\tau_2)$, *for all* $\tau_1, \tau_2 : A \to H$ *with* $\tau_1 \leq_{K^*} \tau_2$.

Proof. Let $\tau_1, \tau_2 : A \to H$ such that $\tau_1 \leq_{K^*} \tau_2$. We have to prove that

$$\langle \mathcal{I}(\tau_1)(x) - \mathcal{I}(\tau_2)(x), y \rangle \geq 0, \tag{3.3}$$

for all $x \in A$ and $y \in K$. For $x = 0$ the inequality is trivial. Suppose that $x \neq 0$. Because $A \backslash \{0\}$ is an invariant set of i, $i(x) \in A$. By the inequality $\tau_1 \leq_{K^*} \tau_2$, we have

$$\langle \tau_1(i(x)) - \tau_2(i(x)), y \rangle \geq 0. \tag{3.4}$$

Multiplying inequality (3.4) by $\|x\|^2$, we obtain the required inequality (3.3). $\square$

We remark that it is easy to see that $\mathcal{I}$ is also K-monotone on Υ.

PROPOSITION 3.5 *If there exist $a, b \in \mathbb{R}$ with $0 < a \leq b$ and $q : K \rightarrow K$ completely continuous with $q \leq_K bI$, such that $f = bI - q$ and*

$$aI \leq_{K^*} f, \tag{3.5}$$

for all $x \in K$, then f is surjective.

Proof. We use Corollary 3.35. The first assumption of Corollary 3.33 is obviously satisfied. It remains to prove that $\underline{\mathcal{I}(f)^{\#}(0)} > 0$. By inequality (3.5) and Lemma 3.36 with $A = K$, we have

$$ax \leq_{K^*} \mathcal{I}(f)(x), \tag{3.6}$$

for all $x \in K \backslash \{0\}$. Because $K \backslash \{0\}$ is invariant under i, we also have $i(x) \in K$. Hence, multiplying scalarly inequality (3.6) by $i(x)$, we obtain

$$\langle \mathcal{I}(f)(x), i(x) \rangle \geq a. \tag{3.7}$$

Tending with x to 0 in (3.7) it yields

$$\underline{\mathcal{I}(f)^{\#}(0)} \geq a > 0. \quad \square$$

$\square$

COROLLARY 3.37 *Consider the case when $H = \mathbb{R}^n$ and $K = \mathbb{R}^n_+$, where*

$$\mathbb{R}^n_+ = \{x = (x_1, \ldots, x_n) \mid x_i \geq 0 \text{ for all } i = 1, \ldots, n\}$$

is the nonnegative orthant of $\mathbb{R}^n$. If f is continuous and there exist $a, b \in \mathbb{R}$, such that $0 < a \leq b$ and

$$aI \leq_K f \leq_K bI, \tag{3.8}$$

then f is surjective.

Proof. It is easy to see that $K = K^*$. Hence, Corollary 3.37 is a straightforward consequence of Proposition 3.5. $\square$

We remark that Corollary 3.37 remains true for generating closed pointed convex cones in the nonnegative orthant and their images through orthogonal transformations (i.e., nonsingular linear transformations of $\mathbb{R}^n$ whose inverse is equal to their transpose). For these cones we have $K \subseteq K^*$ and therefore we can apply Proposition 3.5.

EXAMPLE 3.38 *Let* $H = \mathbb{R}^2$, $K = \mathbb{R}^2_+$, $a, b \in \mathbb{R}$; $0 < a \le b$ *and* α, β : $\mathbb{R}^2_+ \to [a, b]$ *two arbitrary continuous mappings. Define* $f : K \to K$ *by the relation*

$$f(x_1, x_2) = (\alpha(x_1, x_2)x_1, \beta(x_1, x_2)x_2)$$

for every $x = (x_1, x_2) \in \mathbb{R}^2_+$. *It is easy to see that the conditions of Corollary 3.37 are satisfied. Hence,* f *is surjective.*

3.5 Variational Inequalities and Complementarity Problems

Let $(E, \|\cdot\|)$ be a Banach space, E^* the topological dual of E, $\langle E, E^* \rangle$ a duality between E, and E^* and $\langle \cdot, \cdot \rangle$ the bilinear mapping which defines the duality $\langle E, E^* \rangle$.

LEMMA 3.39 *If* $\{x_n\}_{n\in\mathbb{N}} \subseteq E$, $\{y_n\}_{n\in\mathbb{N}} \subseteq E^*$ *are sequences such that* $\{x_n\}_{n\in\mathbb{N}}$ *is weakly convergent to* $x_* \in E$ *and* $\{y_n\}_{n\in\mathbb{N}}$ *is strongly convergent to* $y_* \in E^*$, *then* $\lim_{n\to\infty} \langle x_n, y_n \rangle = \langle x_*, y_* \rangle$.

Proof. The lemma is a consequence of the following formula:

$$\langle x_n, y_n \rangle - \langle x_*, y_* \rangle = \langle x_n - x_*, y_n - y_* \rangle + \langle x_*, y_n \rangle + \langle x_n, y_* \rangle - 2\langle x_*, y_* \rangle.$$

We recall the following classical results.

THEOREM 3.40 [Eberlein–Šmulian] *A set* $M \subseteq E$ *is relatively weakly compact iff every sequence* $\{x_n\}_{n\in\mathbb{N}}$ *in* M *has a weakly convergent subsequence.*

Proof. For a proof of this theorem the reader is referred to Wojtaszczyk [1991]. □

PROPOSITION 3.6 *Any closed ball in* E^* *is* $\sigma(E^*, E)$-*compact.*

Proof. This proposition is Proposition 1 in [Bourbaki, 1964], Chapter IV, p. 112. □

Recall the following definition [Isac and Gowda, 1993].

DEFINITION 3.41 *We say that a mapping* $T_1 : E \to E^*$ *satisfies condition* $(S)^1_+$ *if any sequence* $\{x_n\}_{n\in\mathbb{N}} \subseteq E$ *with the following properties,*

1. $\{x_n\}_{n\in\mathbb{N}}$ *is* $\sigma(E, E^*)$-*convergent to* $x_* \in E$;

2. $\{T_1(x_n)\}_{n \in \mathbb{N}}$ is $\sigma(E^*, E)$-convergent to $u_* \in E^*$;

3. $\limsup_{n \to \infty} \langle x_n, T_1(x_n) \rangle \leq \langle x_*, u_* \rangle$;

has a subsequence convergent to x_*.

REMARK 3.3 *Examples of mappings satisfying condition* $(S)_+^1$ *are given in [Isac and Gowda, 1993].*

DEFINITION 3.42 *We say that a mapping* $T_2 : E \to E^*$ *is demicompletely continuous if the following conditions are satisfied,*

1. T_2 *is continuous.*

2. *For every weakly convergent sequence* $\{x_n\}_{n \in \mathbb{N}} \subseteq E$, *a strongly convergent subsequence exists in* $\{T_2(x_n)\}_{n \in \mathbb{N}}$.

REMARK 3.4 *If E is a reflexive Banach space, then demicomplete continuity and complete continuity are equivalent. However, if E is a nonreflexive Banach space, then this fact is not true.*

In this section we give some applications for variational inequalities and in particular for complementarity problems.

Given a mapping $f : E \to E^*$ and a closed convex set $D \subseteq E$ the *variational inequality* defined by f and D is the following problem,

$$\text{VI}(f, D) : \begin{cases} \text{find } x_* \in D \text{ such that} \\ \langle f(x_*), x - x_* \rangle \geq 0, \text{ for all } x \in D. \end{cases} \tag{3.9}$$

If in particular the set $D = K$ where K is a closed convex cone in E, and the dual cone of K is K^*, then in this case it is known [Isac, 1992, 2000d] that the problem VI(f, K) is equivalent to the following *nonlinear complementarity problem*

$$\text{NCP}(f, K) : \begin{cases} \text{find } x_* \in K \text{ such that} \\ f(x_*) \in K^* \text{ and } \langle x_*, f(x_*) \rangle = 0. \end{cases}$$

The theory of variational inequalities is one of the most popular domains of applied mathematics [Baiocchi and Capello, 1984] and [Kinderlehrer and Stampacchia, 1980].

Complementarity theory is a relatively new domain of applied mathematics with many application in economics, optimization, game theory, mechanics, engineering among others [Cottle et al., 1992; Isac, 1992, 2000d; Isac and Gowda, 1993].

THEOREM 3.43 *Let* $T_1, T_2 : E \to E^*$ *be two mappings. If the following assumptions are satisfied,*

1. T_1 *is continuous, bounded (i.e., for any bounded set $B \subseteq E$, $T(B)$ is bounded) and satisfies condition $(S)^1_+$;*

2. T_2 *is demicompletely continuous;*

then, for every weakly compact nonempty convex set $D \subseteq E$, the variational inequality $\mathrm{VI}(T_1 - T_2, D)$ has a solution.

Proof. Let Λ be the family of all finite-dimensional subspaces F of E such that $F \cap D$ is nonempty. Consider the family Λ ordered by inclusion. Denote by $f(x) = T_1(x) - T_2(x)$ for all $x \in D$ and by $D(F) = F \cap D$, for each $F \in \Lambda$. For each $F \in \Lambda$ we define

$$A_F := \{y \in D \mid \langle x - y, f(y) \rangle \geq 0 \text{ for all } x \in D(F)\}.$$

For each $F \in \Lambda$ the set A_F is nonempty. Indeed, to show this it is sufficient to show that the problem $\mathrm{VI}(f, D(F))$ has a solution (because the solution set of the problem $\mathrm{VI}(f, D(F))$ is a subset of A_F). We show now that the solution set of the problem $\mathrm{VI}(f, D(F))$ is nonempty. Indeed, let $j : F \to E$ denote the inclusion and $j^* : E^* \to F^*$ the adjoint (transpose) of j. By our assumption we have that the mapping

$$j^* \circ f \circ j : D(F) \to F^*$$

is continuous and

$$\langle x - y, (j^* \circ f \circ j)(y) \rangle = \langle j(x - y), (f \circ j)(y) \rangle = \langle x - y, f(y) \rangle,$$

for all $x, y \in D(F)$. Applying the classical Hartman–Stampacchia theorem [Isac, 1992] to the mapping $j^* \circ f \circ j$ and the set $D(F)$ we obtain that the problem $\mathrm{VI}(f, D(F))$ has a solution. So, for any $F \in \Lambda$, the set A_F is nonempty. Denote by $\overline{A}^\sigma_F$ the weak closure of A_F. We have that $\bigcap_{F \in \Lambda} \overline{A}^\sigma_F \neq 0$. Indeed, let $\overline{A}^\sigma_{F_1}, \overline{A}^\sigma_{F_2}, \ldots, \overline{A}^\sigma_{F_n}$ be a finite subfamily of the family $\{\overline{A}^\sigma_F\}_{F \in \Lambda}$. Let F_0 be the finite-dimensional subspace in E generated by $F_1, F_2, \ldots, F_n$. Because $F_k \subseteq F_0$ for all $k = 1, 2, \ldots, n$, we have that $D(F_k) \subseteq D(F_0)$ for all $k = 1, 2, \ldots, n$. We have $A_{F_0} \subseteq A_{F_k}$, which implies $\overline{A}^\sigma_{F_0} \subseteq \overline{A}^\sigma_{F_k}$ for all $k = 1, 2, \ldots, n$, and finally we have that $\bigcap_{k=1}^n \overline{A}^\sigma_{F_k} \neq 0$. Because D is weakly compact we conclude that $\bigcap_{F \in \Lambda} \overline{A}^\sigma_F \neq 0$. Let $y_* \in \bigcap_{F \in \Lambda} \overline{A}^\sigma_F$; that is, for every $F \in \Lambda$, $y_* \in \overline{A}^\sigma_F$. Let $x \in D$ be an arbitrary element. There exists some $F \in \Lambda$ such that $x, y_* \in F$. Because $y_* \in \overline{A}^\sigma_F$, there exists a net $\{y_j\} \subseteq A_F$ such that $\{y_j\}$ is weakly convergent to y_*. By Theorem 3.40 (Eberlein–Šmulian), we can suppose that the net $\{y_j\}$ is a sequence $\{y_n\}_{n \in \mathbb{N}}$ weakly convergent to y_*. We have

$$\begin{cases} \langle y_* - y_n, f(y_n) \rangle \geq 0 \quad \text{and} \\ \langle x - y_n, f(y_n) \rangle \geq 0, \end{cases}$$

or

$$\langle y_n - y_*, T_1(y_n) \rangle \leq \langle y_n - y_*, T_2(y_n) \rangle \qquad (3.10)$$

and

$$\langle x - y_n, T_1(y_n) \rangle \geq \langle x - y_n, T_2(y_n) \rangle. \qquad (3.11)$$

By assumption 2. there exists a subsequence of $\{T_2(y_n)\}_{n \in \mathbb{N}}$, denoted again $\{T_2(y_n)\}_{n \in \mathbb{N}}$, strongly convergent to an element $u_0 \in E^*$. From formula (3.10) and considering Lemma 3.39 we have

$$\limsup_{n \to \infty} \langle y_n - y_*, T_1(y_n) \rangle \leq 0. \qquad (3.12)$$

Because T_1 is bounded and considering Proposition 3.6, we can suppose (taking eventually a subsequence of $\{y_n\}_{n \in \mathbb{N}}$) that $\{T_1(y_n)\}_{n \in \mathbb{N}}$ is weakly convergent to an element $v_0 \in E^*$. Because

$$\langle y_n, T_1(y_n) \rangle = \langle y_n - y_*, T_1(y_n) \rangle + \langle y_*, T_1(y_n) \rangle,$$

and considering formula (3.12) we obtain

$$\limsup_{n \to \infty} \langle y_n, T_1(y_n) \rangle \leq \langle y_*, v_0 \rangle.$$

Hence by condition $(S)_+^1$ we obtain that the sequence $\{y_n\}_{n \in \mathbb{N}}$ has a subsequence, denoted again by $\{y_n\}_{n \in \mathbb{N}}$, strongly convergent to y_*. By assumption 2 we must have $T_2(y_*) = u_0$. From inequality (3.11) we obtain $\langle x - y_*, T_1(y_*) - T_2(y_*) \rangle \geq 0$ for all $x \in D$, and the proof is complete. $\qquad \square$

For every $n \in \mathbb{N}$, we denote

$$B(0, n) = \{x \in E \mid \|x\| \leq n\}.$$

DEFINITION 3.44 *We say that a nonempty subset K of E is a weakly Lindelöf set if the following properties are satisfied.*

1. *K is a closed convex unbounded set.*

2. *For any $n \in \mathbb{N}$ such that $D_n = B(0, n) \cap K$ is non-empty, we have that D_n is a weakly compact set.*

Examples for Lindelöf sets

1. Any closed convex unbounded set in a reflexive Banach space

2. Any closed pointed convex cone with a weakly compact base in an arbitrary Banach space

3. Any closed convex unbounded subset of a closed pointed convex cone K generated by a weakly compact convex set D with $0 \notin D$

THEOREM 3.45 *Let $K \subseteq E$ be a weakly Lindelöf subset and $T_1, T_2 : E \to E^*$ two mappings. If the following assumptions are satisfied,*

1. *T_1 is continuous bounded and satisfies condition $(S)^1_+$;*

2. *T_2 is demicompletely continuous;*

3. *there exists a real number $c > 0$ such that $c \leq \liminf_{\substack{\|x\| \to \infty \\ x \in K}} \langle x, T_1(x) \rangle / \|x\|^2$;*

4. *T_2 has a scalar asymptotic derivative $T'_{2,s,K}(\infty)$ along K such that $\|T'_{2,s,K}(\infty)\| < c$;*

then the problem $\mathrm{VI}(T_1 - T_2, K)$ has a solution.

Proof. We may suppose that for any $n \in \mathbb{N}$, $D_n = B(0, n) \cap K$ is nonempty. We have $K = \bigcup_{n=1}^{\infty} D_n$. For each $n \in \mathbb{N}$, D_n is weakly compact and convex. By Theorem 3.43 the problem $\mathrm{VI}(T_1 - T_2, D_n)$ has a solution $y_n \in D_n$ for every $n \in \mathbb{N}$. Therefore we have

$$\langle x - y_n, (T_1 - T_2)(y_n) \rangle \geq 0 \quad \text{for all } x \in D_n. \tag{3.13}$$

If in (3.13) we put $x = 0$, we obtain

$$\langle y_n, T_1(y_n) \rangle \leq \langle y_n, T_2(y_n) \rangle.$$

The sequence $\{y_n\}_{n \in \mathbb{N}}$ is bounded. Indeed, if we suppose that $\|y_n\| \to \infty$ as $n \to \infty$, then by assumptions 3 and 4 we have (supposing that $\|y_n\| \neq 0$ for all $n \in \mathbb{N}$),

$$c \leq \liminf_{\|y_n\| \to \infty} \frac{\langle y_n, T_1(y_n) \rangle}{\|y_n\|^2} \leq \liminf_{\|y_n\| \to \infty} \frac{\langle y_n, T_2(y_n) \rangle}{\|y_n\|^2} \leq$$
$$\leq \limsup_{\|y_n\| \to \infty} \frac{\langle y_n, T_2(y_n) - T_{2,s}(\infty)(y_n) \rangle}{\|y_n\|^2}$$
$$+ \limsup_{\|y_n\| \to \infty} \frac{\langle y_n, T_{2,s}(\infty)(y_n) \rangle}{\|y_n\|^2} \leq \|T_{2,s}(\infty)\|^2 < c,$$

which is a contradiction. Therefore we conclude that $\{y_n\}_{n \in \mathbb{N}}$ is a bounded sequence. Hence, there exists $m \in \mathbb{N}$ such that $\{y_n\} \subseteq D_m$. Because D_m is weakly compact, by Theorem 3.40 (Eberlein–Šmulian), we have that $\{y_n\}_{n \in \mathbb{N}}$ has a subsequence, denoted again $\{y_n\}_{n \in \mathbb{N}}$, weakly convergent to an element $y_* \in K$. Because T_1 is bounded, by Proposition 3.6, and considering eventually again a subsequence, we can suppose that $\{T_1(y_n)\}_{n \in \mathbb{N}}$ is weakly convergent in E^* (i.e., $\sigma(E^*, E)$-convergent) to an element $u \in E^*$. Let $x \in K$ be an arbitrary element. There exists $n_0 \in \mathbb{N}$ such that $n_0 > m$ and $\{y_*, x\} \subseteq D_{n_0}$, and obviously, $\{y_*, x\} \subseteq D_n$ for all $n \geq n_0$. From formula (3.13) we deduce

$$\langle y_* - y_n, (T_1 - T_2)(y_n) \rangle \geq 0, \tag{3.14}$$

and

$$\langle x - y_n, (T_1 - T_2)(y_n) \rangle \geq 0. \tag{3.15}$$

Because there exists a subsequence $\{T_2(y_{n_k})\}_{k \in \mathbb{N}}$ in $\{T_2(y_n)\}_{n \in \mathbb{N}}$ strongly convergent to an element $w \in E^*$ and inasmuch as

$$\langle y_* - y_{n_k}, T_2(y_{n_k}) \rangle = \langle y_* - y_{n_k}, T_2(y_{n_k}) - w \rangle + \langle y_* - y_{n_k}, w \rangle,$$

by using Lemma 3.39 we obtain that

$$\langle y_* - y_{n_k}, T_2(y_{n_k}) \rangle \to 0 \text{ as } k \to \infty.$$

Therefore, by using (3.14) we have

$$\limsup_{k \to \infty} \langle y_{n_k} - y_*, T_1(y_{n_k}) \rangle \leq \limsup_{k \to \infty} \langle y_{n_k} - y_*, T_2(y_{n_k}) \rangle = 0.$$

From the last inequality and the equality

$$\langle y_{n_k}, T_1(y_{n_k}) \rangle = \langle y_{n_k} - y_*, T_1(y_{n_k}) \rangle + \langle y_*, T_1(y_{n_k}) \rangle,$$

we deduce the inequality

$$\limsup_{k \to \infty} \langle y_{n_k}, T_1(y_{n_k}) \rangle \leq \langle y_*, u \rangle.$$

Because T_1 satisfies condition $(S)_+^1$, we obtain that $\{y_{n_k}\}_{k \in \mathbb{N}}$ contains a subsequence, denoted again $\{y_{n_k}\}_{k \in \mathbb{N}}$, strongly convergent to an element, which obviously must be y_*. Now computing the limit in (3.15), considering the properties of T_1 and T_2, and applying again Lemma 3.39, we obtain that

$$\langle x - y_*, (T_1 - T_2)(y_*) \rangle \geq 0 \quad \text{for all } x \in K;$$

that is, the problem $\text{VI}(T_1 - T_2, K)$ has a solution. □

COROLLARY 3.46 *If either E is a reflexive Banach space and $K \subseteq E$ is an arbitrary closed convex pointed cone, or E is an arbitrary Banach space and $K \subseteq E$ is a closed convex pointed cone with a weakly compact base, and the assumptions 1–4 of Theorem 3.45 are satisfied, then the problem $\text{NCP}(T_1 - T_2, K)$ has a solution.*

Let $(H, \langle \cdot, \cdot \rangle)$ be a Hilbert space.

THEOREM 3.47 *Let $K \in H$ be a closed convex unbounded set such that $K \backslash \{0\}$ is an invariant set of the inversion i and $T_1, T_2 : H \to H$ two mappings. If the assumptions*

1. *T_1 is continuous bounded and satisfies condition $(S)_+^1$;*

2. T_2 *is completely continuous;*

3. *there exists a real number* $c > 0$ *such that* $c \leq \mathcal{I}(T_1)^{\#,K}(0)$;

4. $\overline{\mathcal{I}(T_2)}^{\#,K}(0) < c$;

are satisfied, then the problem $\mathrm{VI}(T_1 - T_2, K)$ *has a solution.*

Proof. Because $K \in H$ is unbounded, closed, and $K \backslash \{0\}$ is an invariant set of i, $0 \in K$ and 0 is a nonisolated point of K. Hence, $\mathcal{I}(T_1)^{\#,K}(0)$ and $\overline{\mathcal{I}(T_2)}^{\#,K}(0)$ are well defined. The proof of Theorem 3.47 follows by Theorem 3.45, from using Lemma 3.28 and a similar argument to the proof of Theorem 3.30.
By Corollary 3.46 and 3.47 we have as follows.

COROLLARY 3.48 *If* $K \subseteq H$ *is a closed pointed convex cone and the assumptions 1–4 of Theorem 3.47 are satisfied, then the problem* $\mathrm{NCP}(T_1 - T_2, K)$ *has a solution.*

3.6 Duality in Nonlinear Complementarity Theory

In 1995 Isac introduced a new topological method in complementarity theory. This method is based on the notion of the exceptional family of elements (EFE), which is related to the topological degree and to the Leray–Schauder alternative.

The notion of EFE was presented in a talk given at the Institute of Applied Mathematics of Academia SINICA (Beijing, China). This new topological method was published in 1997 [Isac et al., 1997]. Since that time many papers, based on this method have been published [Bulavski et al., 1998, 2001; Carbone and Zabreiko, 2002; Hyers et al., 1997; Isac 1999a,b, 2000b,c,d, 2001; Isac and Carbone, 1999; Isac and Cojocaru, 2002; Isac and Kalashnikov, 2001; Isac and Obuchowska, 1998; Isac and Zhao, 2000; Isac et al., 1997, 2001, 2002; Kalashnikov and Isac, 2002; Obuchowska, 2001; Zhao, 1998, 1999; Zhao and Han, 1999; Zhao and Isac, 2000a,b; Zhao and Li, 2000, 2001a,b; Zhao and Sun, 2001; Zhao and Yuan, 2000; Zhao et al., 1999]. The main result presented in [Isac et al., 1997] is the following theorem. If $(H, \langle \cdot, \cdot \rangle)$ is a Hilbert space, $K \subset H$ a closed convex cone and $f : H \to H$ is a completely continuous field, then either the complementarity problem defined by K and f has a solution, or f is without EFE.

This theorem shows that it is very important to know when a given mapping is without EFE. Several classes of mappings with this property were presented in the above-mentioned references.

We note that for a mapping the nonexistence of (EFE) is a kind of very general coercivity condition. In this section we present the notion of "infinitesimal exceptional family of element (IEFE)". This notion was defined here by Németh.

If we consider the couple (EFE, IEFE) we remark a kind of duality, through a special inversion function. By this duality we put in evidence new classes of mappings for which the complementarity problem has a solution. Some interesting relations between the solvability of the complementarity problem and the scalar derivative are also established. The scalar derivative was introduced in [Nemeth, 1992] and studied in several papers, for example [Isac and Nemeth, 2003; Nemeth, 1992, 1993], and [Nemeth, 2006] among others.

This section could open a challenging new research direction in complementarity theory.

3.6.1   Preliminaries

A completely continuous field on H is a mapping $f : H \to H$ such that $f = I - T$, where I is the identical mapping of H (i.e., $I(x) = x$, for all $x \in H$) and T is a completely continuous mapping. In the particular case $H = \mathbb{R}^n$, any continuous mapping $f : \mathbb{R}^n \to \mathbb{R}^n$ is a completely continuous field, because $f = I - (I - f)$.

3.6.2   Complementarity Problem

DEFINITION 3.49 *Let* $(H, \langle \cdot, \cdot \rangle)$ *be a Hilbert space,* $K \subset H$ *a closed convex cone,* K^* *its dual cone, and* $f : K \to H$ *a mapping. The* nonlinear complementarity problem *defined by f and the cone K is*

$$\text{NCP}(f, K) : \begin{cases} \text{find } x_* \in K \text{ such that} \\ f(x_*) \in K^* \text{ and } \langle x_*, f(x_*) \rangle = 0. \end{cases}$$

3.6.3   Exceptional Family of Elements

The next definition can be found in [Isac and Carbone, 1999], and [Isac et al., 1997].

DEFINITION 3.50 *Let* $(H, \langle \cdot, \cdot \rangle)$ *be a Hilbert space,* $K \subset H$ *a closed convex cone, and* $f : H \to H$ *a mapping. We say that a family of elements* $\{x_r\}_{r>0} \subset K$ *is an* exceptional family of elements *for f with respect to K, if for every real number $r > 0$, there exists a real number $\mu_r > 0$ such that the vector* $u_r = \mu_r x_r + f(x_r)$ *satisfies the following conditions;*

1. $u_r \in K^*$;

2. $\langle u_r, x_r \rangle = 0$;

3. $\|x_r\| \to +\infty$ *as* $r \to +\infty$.

The next theorem is Theorem 9 of [Isac, 2001].

THEOREM 3.51 *Let* $(H, \langle \cdot, \cdot \rangle)$ *be a Hilbert space,* $K \subset H$ *a closed convex cone, and* $f : H \to H$ *a completely continuous field. If* f *is without an exceptional family of elements with respect to* K, *then the problem* $\text{NCP}(f, K)$ *has a solution.*

The next definition can be found in [Isac, 1999a] and [Isac and Carbone, 1999].

DEFINITION 3.52 *Let* $(H, \langle \cdot, \cdot \rangle)$ *be a Hilbert space,* $K \subset H$ *a closed convex cone, and* $f : H \to H$ *a mapping. We say that the mapping* f *satisfies* condition Θ *with respect to* K *if*

$$
\begin{cases}
\text{There exists } \rho > 0 \text{ such that for each } x \in K \text{ with } \|x\| > \rho, \\
\text{There exists } p \in K \text{ with } \|p\| < \|x\| \text{ such that} \\
\langle x - p, f(x) \rangle \geq 0.
\end{cases} \tag{3.16}
$$

The next definition is a particular case of condition Θ_g of [Kalashnikov and Isac, 2002] with $g = I$.

DEFINITION 3.53 *Let* $(H, \langle \cdot, \cdot \rangle)$ *be a Hilbert space,* $K \subset H$ *a closed convex cone, and* $f : H \to H$ *a mapping. We say that the mapping* f *satisfies* condition $\tilde{\Theta}$ *with respect to* K *if*

$$
\begin{cases}
\text{There exists } \rho > 0 \text{ such that for each } x \in K \text{ with } \|x\| > \rho, \\
\text{There exists } p \in K \text{ with } \langle p, x \rangle < \|x\|^2 \text{ such that} \\
\langle x - p, f(x) \rangle \geq 0.
\end{cases} \tag{3.17}
$$

The next lemma shows that condition $\tilde{\Theta}$ is an extension of condition Θ.

LEMMA 3.54 *Let* $(H, \langle \cdot, \cdot \rangle)$ *be a Hilbert space,* $K \subset H$ *a closed convex cone, and* $f : H \to H$ *a mapping. If* f *satisfies condition* Θ *with respect to* K, *then it satisfies condition* $\tilde{\Theta}$ *with respect to* K.

Proof. Because f satisfies condition Θ with respect to K, there exists $\rho > 0$ such that for each $x \in K$ with $\|x\| > \rho$, there exists $p \in K$ with $\|p\| < \|x\|$ such that $\langle x - p, f(x) \rangle \geq 0$. By the Cauchy inequality

$$
\langle p, x \rangle \leq \|p\| \|x\| < \|x\|^2.
$$

Hence, f satisfies condition $\tilde{\Theta}$ with respect to K. $\qquad\square$

The next theorem is proved in [Isac, 1999a]. It also follows from Lemma 3.54 and Theorem 3.56.

THEOREM 3.55 *Let* H *be a Hilbert space,* $K \subset H$ *a closed convex cone, and* $f : H \to H$ *a mapping. If* f *satisfies condition* Θ *with respect to* K, *then it is without an exceptional family of elements with respect to* K.

The next result follows from the proof of Theorem 4 [Kalashnikov and Isac, 2002].

THEOREM 3.56 *Let H be a Hilbert space, $K \subset H$ a closed convex cone, and $f : H \to H$ a mapping. If f satisfies condition $\tilde{\Theta}$ with respect to K, then it is without an exceptional family of elements with respect to K.*

3.6.4   Infinitesimal Exceptional Family of Elements

DEFINITION 3.57 *Let $(H, \langle \cdot, \cdot \rangle)$ be a Hilbert space, $K \subset H$ a closed convex cone, and $g : K \to H$ a mapping. We say that $\{y_r\}_{r>0} \subset K$ is an infinitesimal exceptional family of elements for g with respect to K, if for every real number $r > 0$, there exists a real number $\mu_r > 0$ such that the vector $v_r = \mu_r y_r + g(y_r)$ satisfies the following conditions.*

1. $v_r \in K^*$.

2. $\langle v_r, y_r \rangle = 0$.

3. $y_r \to 0$ as $r \to +\infty$.

DEFINITION 3.58 *Let $(H, \langle \cdot, \cdot \rangle)$ be a Hilbert space, $K \subset H$ a closed convex cone, and $g : H \to H$ a mapping. We say that the mapping g satisfies condition ${}^i\Theta$ with respect to K if*

$$\begin{cases} \text{There exists } \lambda > 0 \text{ such that for each } y \in K \backslash \{0\} \text{ with } \|y\| < \lambda, \\ \text{There exists } q \in K \text{ with } \|q\| < \|y\| \text{ such that} \\ \langle y - q, g(y) \rangle \geq 0. \end{cases} \tag{3.18}$$

DEFINITION 3.59 *Let $(H, \langle \cdot, \cdot \rangle)$ be a Hilbert space, $K \subset H$ a closed convex cone, and $g : H \to H$ a mapping. We say that the mapping g satisfies condition ${}^i\tilde{\Theta}$ with respect to K if*

$$\begin{cases} \text{There exists } \lambda > 0 \text{ such that for each } y \in K \backslash \{0\} \text{ with } \|y\| < \lambda, \\ \text{There exists } q \in K \text{ with } \langle q, y \rangle < \|y\|^2 \text{ such that} \\ \langle y - q, g(y) \rangle \geq 0. \end{cases} \tag{3.19}$$

The next lemma shows that condition ${}^i\tilde{\Theta}$ is an extension of condition ${}^i\Theta$ and it can be proved similarly to Lemma 3.54.

LEMMA 3.60 *Let $(H, \langle \cdot, \cdot \rangle)$ be a Hilbert space, $K \subset H$ a closed convex cone, and $g : H \to H$ a mapping. If g satisfies condition ${}^i\Theta$ with respect to K, then it satisfies condition ${}^i\tilde{\Theta}$ with respect to K.*

THEOREM 3.61 *Let $(H, \langle \cdot, \cdot \rangle)$ be a Hilbert space, $K \subset H$ a closed convex cone, and $g : H \to H$ a mapping. If g satisfies condition ${}^i\tilde{\Theta}$ with respect to K,* *then it is without an infinitesimal exceptional family of elements with respect to K.*

Proof. Suppose to the contrary, that g has an infinitesimal family of elements $\{y_r\}_{r>0} \subset K$ with respect to K. For any $r > 0$ such that $\|y_r\| < \rho$ there is an element $q_r \in K$ with $\langle q_r, y_r \rangle < \|y_r\|^2$ satisfying relation (3.18); that is,

$$\langle y_r - q_r, g(y_r) \rangle \geq 0.$$

According to Definition 3.57, $\langle v_r, y_r \rangle = 0$ and $v_r \in K^*$, therefore we have

$$
\begin{aligned}
0 \leq \langle y_r - q_r, g(y_r) \rangle &= \langle y_r - q_r, v_r - \mu_r y_r \rangle \\
&= -\mu_r \|y_r\|^2 - \langle q_r, v_r \rangle + \mu_r \langle q_r, y_r \rangle \\
&\leq -\mu_r (\|y_r\|^2 - \langle q_r, y_r \rangle) < 0,
\end{aligned}
$$

which is a contradiction. $\qquad\square$

REMARK 3.5 *At first sight Theorem 3.61 seems to be a direct consequence of Theorems 3.64 and 3.66, proved in the next section. However, note that there might be an infinitesimal family of elements of g which contains zero.*

COROLLARY 3.62 *Let $(H, \langle \cdot, \cdot \rangle)$ be a Hilbert space, $K \subset H$ a closed convex cone, and $g : H \to H$ a mapping. If g satisfies condition ${}^i\Theta$ with respect to K, then it is without an infinitesimal exceptional family of elements with respect to K.*

Proof. By Lemma 3.60 g satisfies condition ${}^i\tilde{\Theta}$ with respect to K. Hence, by Theorem 3.61 g is without an infinitesimal exceptional family of elements with respect to K. $\square$

REMARK 3.6 *At first sight Corollary 3.62 seems to be a direct consequence of Theorems 3.64 and 3.67, proved in the next section. However, note that there might be an infinitesimal family of elements of g which contains zero.*

3.6.5 A Duality and Main Results

THEOREM 3.63 *Let $(H, \langle \cdot, \cdot \rangle)$ be a Hilbert space $K \subset H$ a closed convex cone, and $f : K \to H$ a mapping. Then $x_* \neq 0$ is a solution of $\mathrm{NCP}(f, K)$ if and only if y_* is a solution of $\mathrm{NCP}(g, K)$, where $y_* = i(x_*)$ is the inversion of x_* and $g = \mathcal{I}(f)$ is the inversion of f.*

Proof.

$$\langle y_*, \mathcal{I}(f)(y_*) \rangle = \langle y_*, \|y_*\|^2 f(i(y_*)) \rangle.$$

Hence,

$$\langle y_*, \mathcal{I}(f)(y_*) \rangle = \|y_*\|^4 \langle i(y_*), f(i(y_*)) \rangle.$$

Because $i^{-1} = i$, we have

$$\langle y_*, g(y_*) \rangle = \frac{1}{\|x_*\|^4} \langle x_*, f(x_*) \rangle. \tag{3.20}$$

It can be similarly proved that

$$\langle g(y_*), z \rangle = \frac{1}{\|x_*\|^2} \langle f(x_*), z \rangle, \tag{3.21}$$

for every $z \in K$. By using (3.20),

$$\langle x_*, f(x_*) \rangle = 0$$

if and only if

$$\langle y_*, g(y_*) \rangle = 0.$$

By using (3.21), $f(x_*) \in K^*$ if and only if $g(y_*) \in K^*$. $\square$

THEOREM 3.64 *Let $(H, \langle \cdot, \cdot \rangle)$ be a Hilbert space, $K \subset H$ a closed convex cone, and $f : K \to H$ a mapping. $\{x_r\}_{r>0} \subset K \backslash \{0\}$ is an exceptional family of elements for f with respect to K if and only if $\{y_r\}_{r>0} \subset K \backslash \{0\}$ is an infinitesimal exceptional family of elements for g with respect to K, where $y_r = i(x_r)$ and $g = \mathcal{I}(f)$.*

Proof. Bearing in mind the notations of Definition 3.57, we have

$$v_r = \mu_r y_r + \|y_r\|^2 f(i(y_r)).$$

Hence,

$$v_r = \|y_r\|^2(\mu_r i(y_r) + f(i(y_r)).$$

Because $i^{-1} = i$, we have

$$v_r = \frac{1}{\|x_r\|^2}(\mu_r x_r + f(x_r)).$$

Hence,

$$v_r = \frac{1}{\|x_r\|^2} u_r.$$

Therefore,

$$\langle v_r, y_r \rangle = \frac{1}{\|x_r\|^4} \langle u_r, x_r \rangle \tag{3.22}$$

and

$$\langle v_r, z \rangle = \frac{1}{\|x_r\|^2} \langle u_r, z \rangle, \tag{3.23}$$

for every $z \in K$. Because $\|x_r\| \cdot \|y_r\| = 1$, $\|x_r\| \to +\infty$ if and only if $y_r \to 0$. By using (3.22),

$$\langle u_r, x_r \rangle = 0$$

if and only if

$$\langle v_r, y_r \rangle = 0.$$

By using (3.23), $u_r \in K^*$ if and only if $v_r \in K^*$. □

THEOREM 3.65 *Let $(H, \langle \cdot, \cdot \rangle)$ be a Hilbert space, $K \subset H$ a closed convex cone, and $f : K \to H$ a completely continuous field with $f(0) \notin K^*$. If every infinitesimal exceptional family of elements for $g = \mathcal{I}(f)$ with respect to K contains 0, then the nonlinear complementarity problem $\mathrm{NCP}(f, K)$ has a nonzero solution.*

Proof. Because $f(0) \notin K^*$, if $\mathrm{NCP}(f, K)$ has a solution, then this solution is nonzero. By Theorem 3.51, it is enough to prove that f is without an exceptional family of elements with respect to K. Suppose to the contrary that $\{x_r\}_{r>0}$ is an exceptional family of elements for f with respect to K. Because $f(0) \notin K^*$, by the definition of an exceptional family of elements $\{x_r\}_{r>0} \subset K \backslash \{0\}$. Hence, by Theorem 3.64, $g = \mathcal{I}(f)$ has an infinitesimal exceptional family of elements with respect to K which does not contain 0, which is a contradiction. □

THEOREM 3.66 *Let $(H, \langle \cdot, \cdot \rangle)$ be a Hilbert space, $K \subset H$ a closed convex cone, and $f : H \to H$ a mapping and $g = \mathcal{I}(f)$. Then, f satisfies condition $\tilde{\Theta}$ with respect to K if and only if g satisfies condition $^i\tilde{\Theta}$ with respect to K.*

Proof. Suppose that g satisfies condition $^i\tilde{\Theta}$ with respect to K and prove that f satisfies condition $\tilde{\Theta}$ with respect to K. Consider the constant λ of condition $^i\tilde{\Theta}$ and let

$$\rho = \frac{1}{\lambda}.$$

Let $x \in K$ with

$$\|x\| > \rho \tag{3.24}$$

and $y = i(x)$. Inasmuch as

$$\|y\| = \frac{1}{\|x\|},$$

it follows that $\|y\| < \lambda$. Hence, by condition $^i\tilde{\Theta}$, there exists $q \in K$ with $\langle q, y \rangle < \|y\|^2$ such that $\langle y - q, g(y) \rangle \geq 0$. Let

$$p = \frac{q}{\|y\|^2}. \tag{3.25}$$

Because $\langle q, y \rangle < \|y\|^2$ and $i^{-1} = i$, relation (3.25) implies that

$$\langle p, x \rangle = \frac{\langle q, y \rangle}{\|y\|^4} < \frac{1}{\|y\|^2} = \|x\|^2. \qquad (3.26)$$

On the other hand $\mathcal{I}^{-1} = \mathcal{I}$ implies that

$$\langle x - p, f(x) \rangle = \langle x - p, \mathcal{I}(g)(x) \rangle$$
$$= \langle x - p, \|x\|^2 g(i(x)) \rangle = \|x\|^4 \langle y - q, g(y) \rangle \geq 0. \qquad (3.27)$$

By (3.24), (3.26), and (3.27) f satisfies condition $\tilde{\Theta}$ with respect to K. Now, suppose that f satisfies condition $\tilde{\Theta}$ with respect to K and prove that g satisfies condition $^{i}\tilde{\Theta}$ with respect to K. Consider the constant $\rho > 0$ of condition $\tilde{\Theta}$ and let

$$\lambda = \frac{1}{\rho}.$$

Let $y \in K \backslash \{0\}$ with $\|y\| < \lambda$. We have to prove that there exists $q \in K$ with $\langle q, y \rangle < \|y\|^2$ such that $\langle y - q, g(y) \rangle \geq 0$. Because $f = \mathcal{I}(g)$, we can proceed as above. $\square$

The next theorem can be proved similarly to Theorem 3.66.

THEOREM 3.67 *Let* $(H, \langle \cdot, \cdot \rangle)$ *be a Hilbert space,* $K \subset H$ *a closed convex cone,* $f : H \to H$ *a mapping, and* $g = \mathcal{I}(f)$. *Then,* f *satisfies condition* Θ *with respect to* K *if and only if* g *satisfies condition* $^{i}\tilde{\Theta}$ *with respect to* K.

THEOREM 3.68 *Let* H *be a Hilbert space,* $K \subset H$ *a closed convex cone, and* $f : K \to H$ *a completely continuous field. If* $g = \mathcal{I}(f)$ *satisfies condition* $^{i}\tilde{\Theta}$ *with respect to* K, *then the nonlinear complementarity problem* NCP(f, K) *has a solution.*

Proof. By Theorem 3.67, f satisfies condition Θ with respect to K. Hence, Theorems 3.55 and 3.51 imply that the nonlinear complementarity problem NCP(f, K) has a solution. $\square$

THEOREM 3.69 *Let* H *be a Hilbert space,* $K \subset H$ *a closed convex cone, and* $f : K \to H$ *a completely continuous field. If* $g = \mathcal{I}(f)$ *satisfies condition* $^{i}\tilde{\Theta}$ *with respect to* K, *then the nonlinear complementarity problem* NCP(f, K) *has a solution.*

Proof. By Theorem 3.66, f satisfies condition $\tilde{\Theta}$ with respect to K. Hence, Theorem 4 of [Kalashnikov and Isac, 2002] implies that the nonlinear complementarity problem NCP(f, K) has a solution. $\square$

THEOREM 3.70 *Let* $(H, \langle \cdot, \cdot \rangle)$ *be a Hilbert space,* $K \subset H$ *a closed convex cone, and* $f : K \to H$ *a completely continuous field. If there is a* $\delta > 0$ *and a* $h : B(0, \delta) \cap K \to H$ *with* $h(0) = 0$ *and*

$$\begin{cases} \overline{h}^{\#}(0) < 1, \\ (I - h, \mathcal{I}(f))^{\#}(0) > 0, \end{cases}$$

where $B(0, \delta) = \{z \in H : \|z\| < \delta\}$, *then the nonlinear complementarity problem* $\text{NCP}(f, K)$ *has a solution.*

Proof. Let $g = \mathcal{I}(f)$. Because $\overline{h}^{\#}(0) < 1$, there is a λ_1 with $0 < \lambda_1 < \delta$ such that for every $y \in K$ with $\|y\| < \lambda_1$ we have

$$\langle h(y), y \rangle < \|y\|^2. \tag{3.28}$$

Because

$$(I - h, g)^{\#}(0) > 0,$$

there is a λ_2 with $0 < \lambda_2 < \delta$ such that for every $y \in K$ with $\|y\| < \lambda_2$ we have

$$\langle y - h(y), g(y) \rangle > 0. \tag{3.29}$$

Let $\lambda = \min\{\lambda_1, \lambda_2\}$. Obviously,

$$\lambda > 0. \tag{3.30}$$

For

$$\|y\| < \lambda \tag{3.31}$$

let $q = h(y)$. Then, relations (3.28) and (3.29) imply

$$\langle q, y \rangle < \|y\|^2. \tag{3.32}$$

and

$$\langle y - q, g(y) \rangle \geq 0, \tag{3.33}$$

respectively. Hence, relations (3.30) through (3.33) imply that g satisfies condition $^i\tilde{\Theta}$. Hence, Theorem 3.69 implies that the problem $\text{NCP}(f, K)$ has a solution. $\qquad\square$

In the particular case $h = 0$ we have as follows:

COROLLARY 3.71 *Let* $(H, \langle \cdot, \cdot \rangle)$ *be a Hilbert space,* $K \subset H$ *a closed convex cone, and* $f : K \to H$ *a completely continuous field. If* $\mathcal{I}(f)^{\#}(0) > 0$, *then the nonlinear complementarity problem* $\text{NCP}(f, K)$ *has a solution.*

3.7   Duality of Implicit Complementarity Problems

It is known that complementarity theory has many applications in optimization theory, in engineering, in mechanics, in game theory and in economics.

It seems that the implicit complementarity problem was introduced into complementarity theory in [Bensoussan et al., 1973] and [Bensoussan and Lions, 1973] as a mathematical tool in the study of a stochastic optimal control problem. We note that the implicit complementarity problem has been studied by several authors, for example, V. A. Bulavsky, G. Isac, and V. V. Kalashnikov [Bulavski et al., 1998], J. Capuzzo-Dolcetta and U. Mosco [Capuzzo-Dolcetta and Mosco, 1980], G. Isac [Isac, 1986, 1990, 2000a], G. Isac, V. A. Bulavsky, and V. V. Kalashnikov [Isac et al., 1997], G. Isac and D. Goeleven [Isac and Goeleven, 1993a,b], U. Mosco [Mosco, 1976, 1980], and J. S. Pang among others.

Generally, the implicit complementarity problem has been studied via variational or quasi-variational inequalities [Bensoussan and Lions, 1973; Bensoussan et al., 1973], fixed point theory [Isav, 1986, 1990; Isac and Goeleven, 1993a], iterative methods [Isac and Goeleven, 1993b] or via the notion of zero-epi mapping [Isac, 2000a]. Recently, a new topological method in complementarity theory has been introduced by Isac in [Isac et al., 1997] and studied with Bulavsky and Kalashnikov in several papers. This method is based on the concept of the exceptional family of elements for a mapping. This method has been used in many papers [Bulavski et al., 1998, 2001; Carbone and Zabreiko, 2002; Hyers et al., 1997; Isac, 1999a,b, 2000b,c,d, 2001; Isac and Carbone, 1999; Isac and Cojocaru, 2002; Isac and Kalashnikov, 2001; Isac and Obuchowska, 1998; Isac and Zhao, 2000; Isac et al., 1997, 2001, 2002; Kalashnikov and Isac, 2002; Obuchowska, 2001; Zhao, 1998, 1999; Zhao and Han, 1999; Zhao and Isac, 2000a,b; Zhao and Li, 2000, 2001a,b; Zhao and Sun, 2001; Zhao and Yuan, 2000; Zhao et al., 1999]. The notion of exceptional family of elements is the main mathematical tool used in the study of implicit complementarity problems in the recent paper [Kalashnikov and Isac, 2002].

In this section we introduce the notion of the infinitesimal exceptional family of elements for a mapping and a duality between this notion and the notion of the exceptional family of elements. This notion is due to Németh. By this duality and by using a special scalar derivative we obtain new results for implicit complementarity problems. By this method a new research direction in implicit complementarity theory is now opened.

3.7.1   Implicit Complementarity Problem

DEFINITION 3.72 *Let $(H, \langle \cdot, \cdot \rangle)$ be a Hilbert space and $K \subset H$ a closed pointed convex cone. Given two mappings $f, g : H \to H$, the* implicit complementarity problem *defined by the ordered pair of mappings (f, g) and the* *cone K is*

$$\text{ICP}(f, g, K) : \begin{cases} \text{find } x_* \in K \text{ such that} \\ f(x_*) \in K^*, \ g(x_*) \in K \text{ and} \\ \langle g(x_*), f(x_*) \rangle = 0. \end{cases}$$

3.7.2 Exceptional Family of Elements for an Ordered Pair of Mappings

The next definition is Definition 2 of [Kalashnikov and Isac, 2002].

DEFINITION 3.73 *Let* $(H, \langle \cdot, \cdot \rangle)$ *be a Hilbert space,* $K \subset H$ *a closed pointed convex cone, and* $f, g : H \to H$ *two mappings. We say that a family of elements* $\{x_r\}_{r>0}$ *is an* exceptional family of elements (EFE) *for the ordered pair of mappings* (f, g) *with respect to* K, *if the following conditions are satisfied.*

1. $\|x_r\| \to +\infty$ *as* $r \to +\infty$.

2. *For any* $r > 0$, *there exists* $\mu_r > 0$ *such that* $s_r = \mu_r x_r + f(x_r) \in K^*$, $v_r = \mu_r x_r + g(x_r) \in K$, *and* $\langle v_r, s_r \rangle = 0$.

The next theorem is Theorem 3 of [Kalashnikov and Isac, 2002].

THEOREM 3.74 *Let* $(H, \langle \cdot, \cdot \rangle)$ *be a Hilbert space,* $K \subset H$ *a closed pointed convex cone, and* $f, g : H \to H$ *completely continuous fields such that* $f(x) = x - T(x)$ *and* $g(x) = x - S(x)$, *where* $T, S : H \to H$ *are completely continuous mappings.*

Then there exists either a solution to the implicit complementarity problem ICP(f, g, K) *defined by* K *and the ordered pair of mappings* (f, g) *or an exceptional family of elements* $\{x\}_{r>0}$ *for* (f, g) *with respect to* K. *Moreover, if* $S(K) \subseteq K$, *we have that the problem* ICP(f, g, K) *has either a solution in* K *or an exceptional family of elements* $\{x\}_{r>0} \subset K$.

The next definition can be found in [Kalashnikov and Isac, 2002].

DEFINITION 3.75 *Let* $(H, \langle \cdot, \cdot \rangle)$ *be a Hilbert space,* $K \subset H$ *a closed pointed convex cone, and* $f, g : H \to H$ *two mappings. We say that the mapping* f *satisfies* condition Θ_g *with respect to* K *if there exists* $\rho > 0$ *such that for any* $x \in K$, $\|x\| > \rho$, *there exists* $y \in K$ *such that*

$$\begin{cases} \langle g(x) - y, f(x) \rangle \geq 0 \quad \text{and} \\ \langle g(x) - y, x \rangle > 0 \end{cases} \tag{3.34}$$

The next result follows from the proof of Theorem 4 in [Kalashnikov and Isac, 2002].

THEOREM 3.76 *Let* H *be a Hilbert space,* $K \subset H$ *a closed pointed convex cone, and* $f, g : H \to H$ *two mappings. If* f *satisfies condition* Θ_g *with respect to* K, *then the ordered pair of mappings* (f, g) *is without EFEs with respect to* K.

3.7.3 Infinitesimal Exceptional Family of Elements for an Ordered Pair of Mappings

DEFINITION 3.77 *Let* $(H, \langle \cdot, \cdot \rangle)$ *be a Hilbert space,* $K \subset H$ *a closed pointed convex cone, and* $\tilde{f}, \tilde{g} : H \to H$ *two mappings. We say that a family of elements* $\{\tilde{x}_r\}_{r>0}$ *is an* infinitesimal exceptional family of elements IEFE *for the ordered pair of mappings* $(\tilde{f}, \tilde{g})$ *with respect to* K, *if the following conditions are satisfied.*

1. $\tilde{x}_r \to 0$ *as* $r \to +\infty$.

2. *For any* $r > 0$, *there exists* $\mu_r > 0$ *such that* $\tilde{s}_r = \mu_r \tilde{x}_r + \tilde{f}(\tilde{x}_r) \in K^*$, $\tilde{v}_r = \mu_r \tilde{x}_r + \tilde{g}(\tilde{x}_r) \in K$, *and* $\langle \tilde{v}_r, \tilde{s}_r \rangle = 0$.

DEFINITION 3.78 *Let* $(H, \langle \cdot, \cdot \rangle)$ *be a Hilbert space,* $K \subset H$ *a closed pointed convex cone, and* $\tilde{f}, \tilde{g} : H \to H$ *two mappings. We say that the mapping* $\tilde{f}$ *satisfies* condition ${}^i\Theta_{\tilde{g}}$ *with respect to* K *if there exists* $\tilde{\rho} > 0$ *such that for each* $\tilde{x} \in K \backslash \{0\}$ *with* $\|\tilde{x}\| < \tilde{\rho}$ *there exists* $\tilde{y} \in K$ *such that*

$$\begin{cases} \langle \tilde{g}(\tilde{x}) - \tilde{y}, \tilde{f}(\tilde{x}) \rangle \geq 0 \quad \text{and} \\ \langle \tilde{g}(\tilde{x}) - \tilde{y}, \tilde{x} \rangle > 0 \end{cases} \qquad (3.35)$$

THEOREM 3.79 *Let* $(H, \langle \cdot, \cdot \rangle)$ *be a Hilbert space,* $K \subset H$ *a closed convex cone, and* $\tilde{f}, \tilde{g} : H \to H$ *two mappings. If* $\tilde{f}$ *satisfies condition* ${}^i\Theta_{\tilde{g}}$ *with respect to* K, *then the ordered pair of mappings* $(\tilde{f}, \tilde{g})$ *is without IEFEs with respect to* K.

Proof. Suppose to the contrary, that $(\tilde{f}, \tilde{g})$ has an infinitesimal family of elements $\{\tilde{x}_r\}_{r>0} \subset K$. For any $r > 0$ such that $\|\tilde{x}_r\| < \rho$ there is an element $\tilde{y}_r \in K$ with satisfying relation (3.35); that is,

$$\begin{cases} \langle \tilde{g}(\tilde{x}_r) - \tilde{y}_r, \tilde{f}(\tilde{x}_r) \rangle \geq 0 \quad \text{and} \\ \langle \tilde{g}(\tilde{x}_r) - \tilde{y}_r, \tilde{x}_r \rangle \geq 0 \end{cases}$$

Because, according to Definition 3.77, $\tilde{s}_r = \mu_r \tilde{x}_r + \tilde{f}(\tilde{x}_r) \in K^*$, $\tilde{v}_r = \mu_r \tilde{x}_r + \tilde{g}(\tilde{x}_r) \in K^*$, and $\langle \tilde{v}_r, \tilde{s}_r \rangle = 0$, we have

$$\begin{aligned} 0 &\leq \langle \tilde{g}(\tilde{x}_r) - \tilde{y}_r, \tilde{f}(\tilde{x}_r) \rangle = \langle \tilde{v}_r - \mu_r \tilde{x}_r - \tilde{y}_r, \tilde{s}_r - \mu_r \tilde{x}_r \rangle \\ &= \langle \tilde{v}_r, \tilde{s}_r \rangle - \langle \mu_r \tilde{x}_r, \tilde{s}_r \rangle - \langle \tilde{y}_r, \tilde{s}_r \rangle - \langle \tilde{v}_r, \mu_r \tilde{x}_r \rangle + \mu_r^2 \|\tilde{x}_r\|^2 + \langle \tilde{y}_r, \mu_r \tilde{x}_r \rangle \\ &\leq -\langle \tilde{v}_r, \mu_r \tilde{x}_r \rangle + \mu_r^2 \|\tilde{x}_r\|^2 + \langle \tilde{y}_r, \mu_r \tilde{x}_r \rangle \\ &= -\langle \mu_r \tilde{x}_r + \tilde{g}(\tilde{x}_r), \mu_r \tilde{x}_r \rangle + \mu_r^2 \|\tilde{x}_r\|^2 + \langle \tilde{y}_r, \mu_r \tilde{x}_r \rangle \\ &= -\mu_r^2 \|\tilde{x}_r\|^2 - \langle \tilde{g}(\tilde{x}_r), \mu_r \tilde{x}_r \rangle + \mu_r^2 \|\tilde{x}_r\|^2 + \langle \tilde{y}_r, \mu_r \tilde{x}_r \rangle \\ &= -\langle \tilde{g}(\tilde{x}_r), \mu_r \tilde{x}_r \rangle + \langle \tilde{y}_r, \mu_r \tilde{x}_r \rangle = -\mu_r \langle \tilde{g}(\tilde{x}_r) - \tilde{y}_r, \tilde{x}_r \rangle < 0, \end{aligned}$$

which is a contradiction. Hence, the pair $(\tilde{f}, \tilde{g})$ is without IEFEs with respect to K. $\square$

3.7.4   A Duality and Main Results

THEOREM 3.80 *Let* $(H, \langle \cdot, \cdot \rangle)$ *be a Hilbert space,* $K \subset H$ *a closed pointed convex cone, and* $f, g : K \to H$ *two mappings. Then* $x_* \neq 0$ *is a solution of* ICP(f, g, K) *if and only if* $\tilde{x}_*$ *is a solution of* ICP$(\tilde{f}, \tilde{g}, K)$, *where* $\tilde{x}_* = i(x_*)$, $\tilde{f} = \mathcal{I}(f)$, *and* $\tilde{g} = \mathcal{I}(g)$.

Proof.

$$\langle \tilde{g}(\tilde{x}_*), \tilde{f}(\tilde{x}_*) \rangle = \langle \|\tilde{x}_*\|^2 g(i(\tilde{x}_*)), \|\tilde{x}_*\|^2 f(i(\tilde{x}_*)) \rangle.$$

Because $\|\tilde{x}_*\| \cdot \|x_*\| = 1$ and $i^{-1} = i$, we have

$$\langle \tilde{g}(\tilde{x}_*), \tilde{f}(\tilde{x}_*) \rangle = \frac{1}{\|x_*\|^4} \langle g(x_*), f(x_*) \rangle. \tag{3.36}$$

It can be similarly proved that

$$\langle \tilde{f}(\tilde{x}_*), z \rangle = \frac{1}{\|x_*\|^2} \langle f(x_*), z \rangle, \tag{3.37}$$

for every $z \in K$. We also have

$$\tilde{g}(\tilde{x}_*) = \frac{1}{\|x_*\|^2} g(x_*). \tag{3.38}$$

By using (3.36), $\langle g(x_*), f(x_*) \rangle = 0$ if and only if $\langle \tilde{g}(\tilde{x}_*), \tilde{f}(\tilde{x}_*) \rangle = 0$. By using (3.37), $f(x_*) \in K^*$ if and only if $\tilde{f}(\tilde{x}_*) \in K^*$. By using (3.38), $g(x_*) \in K$ if and only if $\tilde{g}(\tilde{x}_*) \in K$. $\square$

THEOREM 3.81 *Let* $(H, \langle \cdot, \cdot \rangle)$ *be a Hilbert space,* $K \subset H$ *a closed pointed convex cone, and* $f, g : K \to H$ *two mappings.* $\{x_r\}_{r>0} \subset H \backslash \{0\}$ *is an* EFE *for the ordered pair of mappings* (f, g) *with respect to* K *if and only if* $\{\tilde{x}_r\}_{r>0} \subset H \backslash \{0\}$ *is an* IEFE *for the ordered pair of mappings* $(\tilde{f}, \tilde{g})$ *with respect to* K, *where* $\tilde{x}_r = i(x_r)$, $\tilde{f} = \mathcal{I}(f)$, *and* $\tilde{g} = \mathcal{I}(f)$.

Proof. Bearing in mind the notations of Definition 3.77, we have

$$\tilde{s}_r = \mu_r \tilde{x}_r + \|\tilde{x}_r\|^2 f(i(\tilde{x}_r))$$

and

$$\tilde{v}_r = \mu_r \tilde{x}_r + \|\tilde{x}_r\|^2 g(i(\tilde{x}_r)).$$

Hence,

$$\tilde{s}_r = \|\tilde{x}_r\|^2 (\mu_r i(\tilde{x}_r) + f(i(\tilde{x}_r)))$$

and

$$\tilde{v}_r = \|\tilde{x}_r\|^2 (\mu_r i(\tilde{x}_r) + g(i(\tilde{x}_r))).$$

Because $\|\tilde{x}_*\| \cdot \|x_*\| = 1$ and $i^{-1} = i$, we have,

$$\tilde{s}_r = \frac{1}{\|x_r\|^2}(\mu_r x_r + f(x_r))$$

and

$$\tilde{v}_r = \frac{1}{\|x_r\|^2}(\mu_r x_r + g(x_r)).$$

Hence,

$$\tilde{s}_r = \frac{1}{\|x_r\|^2}s_r$$

and

$$\tilde{v}_r = \frac{1}{\|x_r\|^2}v_r. \tag{3.39}$$

Therefore,

$$\langle \tilde{v}_r, \tilde{s}_r \rangle = \frac{1}{\|x_r\|^4}\langle v_r, s_r \rangle \tag{3.40}$$

and

$$\langle \tilde{s}_r, y \rangle = \frac{1}{\|x_r\|^2}\langle s_r, y \rangle, \tag{3.41}$$

for every $y \in K$. Because $\|x_r\| \cdot \|\tilde{x}_r\| = 1$, $\|x_r\| \to +\infty$ if and only if $\tilde{x}_r \to 0$.
By using (3.40),

$$\langle v_r, s_r \rangle = 0$$

if and only if

$$\langle \tilde{v}_r, \tilde{s}_r \rangle = 0.$$

By using (3.41), $s_r \in K^*$ if and only if $\tilde{s}_r \in K^*$. By using 3.39, $v_r \in K$ if and
only if $\tilde{v}_r \in K$. $\square$

THEOREM 3.82 *Let* $(H, \langle \cdot, \cdot \rangle)$ *be a Hilbert space,* $K \subset H$ *a closed pointed convex cone, and* $f, g : K \to H$ *completely continuous fields with* $(f(0), g(0)) \notin K^* \times K$. *If every infinitesimal exceptional family of elements for the ordered pair of mappings* $(\tilde{f}, \tilde{g}) = (\mathcal{I}(f), \mathcal{I}(g))$ *with respect to* K *contains* 0, *then the implicit complementarity problem* ICP(f, g, K) *has a nonzero solution.*

   Proof. Because $(f(0), g(0)) \notin K^* \times K$, if ICP$(f, g, K)$ has a solution, then this solution is nonzero. By Theorem 3.74, it is enough to prove that the ordered pair of mappings (f, g) is without an exceptional family of elements with respect to K. Suppose to the contrary that $\{x_r\}_{r>0}$ is an exceptional family of elements for (f, g) with respect to K. Because $(f(0), g(0)) \notin K^* \times K$, by the definition of an exceptional family of elements $\{x_r\}_{r>0} \subset H \backslash \{0\}$. Hence, by Theorem 3.81, $(\tilde{f}, \tilde{g})$ has an infinitesimal exceptional family of elements with respect to K which does not contain 0, which is a contradiction. $\square$

THEOREM 3.83 *Let $(H, \langle \cdot, \cdot \rangle)$ be a Hilbert space, $K \subset H$ a closed pointed convex cone, $f, g : H \to H$ two mappings, $\tilde{f} = \mathcal{I}(f)$ and $\tilde{g} = \mathcal{I}(g)$. Then, f satisfies condition Θ_g with respect to K if and only if $\tilde{f}$ satisfies condition $^i\Theta_{\tilde{g}}$ with respect to K.*

Proof. Suppose that $\tilde{f}$ satisfies condition $^i\Theta_{\tilde{g}}$ with respect to K and prove that f satisfies condition Θ_g with respect to K. Consider the constant $\tilde{\rho}$ of condition $^i\Theta_{\tilde{g}}$ and let

$$\rho = \frac{1}{\tilde{\rho}}.$$

Let $x \in K$ with

$$\|x\| > \rho \tag{3.42}$$

and $\tilde{x} = i(x)$. Inasmuch as

$$\|\tilde{x}\| = \frac{1}{\|x\|},$$

it follows that $\|\tilde{x}\| < \tilde{\rho}$. Hence, by condition $^i\Theta_{\tilde{g}}$, there exists $\tilde{y} \in K$ such that

$$\begin{cases} \langle \tilde{g}(\tilde{x}) - \tilde{y}, \tilde{f}(\tilde{x}) \rangle \geq 0 & \text{and} \\ \langle \tilde{g}(\tilde{x}) - \tilde{y}, \tilde{x} \rangle > 0 \end{cases}$$

Let

$$y = \frac{\tilde{y}}{\|\tilde{x}\|^2}. \tag{3.43}$$

Because $\langle \tilde{g}(\tilde{x}) - \tilde{y}, \tilde{x} \rangle > 0$, $i^{-1} = i$ and relation (3.43) implies that

$$\langle g(x) - y, x \rangle = \left\langle g(i(\tilde{x})) - \frac{\tilde{y}}{\|\tilde{x}\|^2}, \frac{\tilde{x}}{\|\tilde{x}\|^2} \right\rangle = \frac{1}{\|\tilde{x}\|^4} \langle \tilde{g}(\tilde{x}) - \tilde{y}, \tilde{x} \rangle > 0. \tag{3.44}$$

On the other hand, $\langle \tilde{g}(\tilde{x}) - \tilde{y}, \tilde{f}(\tilde{x}) \rangle \geq 0$, $i^{-1} = i$ and relation (3.43) implies that

$$\langle g(x) - y, f(x) \rangle = \left\langle g(i(\tilde{x})) - \frac{\tilde{y}}{\|\tilde{x}\|^2}, f(i(\tilde{x})) \right\rangle \tag{3.45}$$

$$= \frac{1}{\|\tilde{x}\|^4} \langle \tilde{g}(\tilde{x}) - \tilde{y}, \tilde{f}(\tilde{x}) \rangle \geq 0.$$

By (3.42), (3.44), and (3.45) f satisfies condition Θ_g with respect to K. Since $\mathcal{I}^{-1} = \mathcal{I}$ the converse can be proved similarly. □

THEOREM 3.84 *Let H be a Hilbert space, $K \subset H$ a closed pointed convex cone, and $f, g : K \to H$ completely continuous fields, $\tilde{f} = \mathcal{I}(f)$ and $\tilde{g} = \mathcal{I}(g)$. If $\tilde{f}$ satisfies condition $^i\Theta_{\tilde{g}}$ with respect to K, then the implicit complementarity problem $\mathrm{ICP}(f, g, K)$ has a solution.*

Proof. By Theorem 3.83, f satisfies condition Θ_g with respect to K. Hence, Theorems 3.74 and 3.76 imply that ICP(f, g, K) has a solution. $\square$

THEOREM 3.85 *Let* $(H, \langle \cdot, \cdot \rangle)$ *be a Hilbert space,* $K \subset H$ *a closed pointed convex cone, and* $f, g : K \to H$ *completely continuous fields. If there is a* $\delta > 0$ *and an* $h : B(0, \delta) \cap K \to H$ *with* $h(0) = 0$ *and*

$$
\begin{cases}
\mathcal{I}(g)^{\#}(0) > \overline{h}^{\#}(0), \\
(\mathcal{I}(g) - h, \tilde{f})^{\#}(0) > 0,
\end{cases}
$$

where $B(0, \delta) = \{z \in H : \|z\| < \delta\}$, *then the implicit complementarity problem ICP(f, g, K) has a solution.*

Proof. Let $\tilde{g} = \mathcal{I}(g)$.

$$
\tilde{g}^{\#}(0) > \overline{h}^{\#}(0),
$$

therefore we have

$$
(\tilde{g} - h)^{\#}(0) > 0.
$$

Hence, there is a λ_1 with $0 < \lambda_1 < \delta$ such that for every $\tilde{x} \in K$ with $\|\tilde{x}\| < \lambda_1$, we have

$$
\langle \tilde{g}(\tilde{x}) - h(\tilde{x}), \tilde{x} \rangle > 0. \tag{3.46}
$$

Because

$$
(\tilde{g} - h, \tilde{f})^{\#}(0) > 0,
$$

there is a λ_2 with $0 < \lambda_2 < \delta$ such that for every $\tilde{x} \in K$ with $\|\tilde{x}\| < \lambda_2$, we have

$$
\langle \tilde{g}(\tilde{x}) - h(\tilde{x}), \tilde{f}(\tilde{x}) \rangle > 0. \tag{3.47}
$$

Let $\tilde{\rho} = \min\{\lambda_1, \lambda_2\}$. Obviously,

$$
\tilde{\rho} > 0. \tag{3.48}
$$

For

$$
\|\tilde{x}\| < \tilde{\rho} \tag{3.49}
$$

let $\tilde{y} = h(\tilde{x})$. Then, relations (3.46) and (3.47) imply

$$
\langle \tilde{g}(\tilde{x}) - \tilde{y}, \tilde{x} \rangle > 0 \tag{3.50}
$$

and

$$
\langle \tilde{g}(\tilde{x}) - \tilde{y}, \tilde{f}(\tilde{x}) \rangle \geq 0, \tag{3.51}
$$

respectively. Hence, relations (3.48) through (3.51) imply that $\tilde{f}$ satisfies condition $^i\tilde{\Theta}$. Hence, Theorem 3.84 implies that the problem ICP(f, g, K) has a solution. $\square$

In the particular case $h = 0$ we have as follows.

COROLLARY 3.86 *Let* $(H, \langle \cdot, \cdot \rangle)$ *be a Hilbert space,* $K \subset H$ *a closed pointed convex cone, and* $f, g : K \to H$ *completely continuous fields. If*

$$\begin{cases} \mathcal{I}(g)^{\#}(0) > 0, \\ \overline{(\mathcal{I}(g), \tilde{f})}^{\#}(0) > 0, \end{cases}$$

then the implicit complementarity problem $ICP(f, g, K)$ *has a solution.*

3.8 Duality of Multivalued Complementarity Problems

The complementarity theory is now in development. The main goal of this relatively new domain of applied mathematics is the study of complementarity problems. In many practical problems the complementarity problems are related to the study of equilibrium as it is considered in physics, in engineering, and also the equilibrium of economic systems.

There exist four classes of complementarity problems: (1) explicit complementarity problems, (2) implicit complementarity problems, (3) complementarity problems with respect to an ordering, and (4) multivalued complementarity problems. The *multivalued complementarity problems* are considered because in many practical problems instead of single-valued mappings set-valued mappings arise. The set-valued mappings are related to the presence of perturbations in the approximate definition of function values or to the uncertainty in mathematical models. Although many results have been obtained for complementarity problems defined by single-valued mappings, there are relatively few papers dedicated to complementarity problems defined by set-valued mappings (see [Chang and Huang, 1991, 1993a,b,c; Gowda and Pang, 1992; Huang, 1998; Hyers et al., 1997; Isac, 1992, 1999b, 2000d; Luna, 1975; Pang, 1995; Parida and Sen, 1987; Saigal, 1976]).

In this section we present several results on multivalued complementarity problems by using the notions of the *exceptional family of elements, infinitesimal exceptional family of elements*, and *scalar derivatives*.

By a special inversion we introduce a *duality* between the *exceptional family of elements* and the *infinitesimal exceptional family of elements*. By using this duality we show how new classes of set-valued mappings for which the multivalued complementarity problem has a solution can be obtained. We used a similar duality in our papers [Isac and Nemeth, 2006a,b] dedicated to the complementarity problems defined by single-valued mappings.

This section emphasizes the effectiveness of the topological method based on the notion of exceptional family of elements introduced by the first author of this section in [Isac et al., 1997] and applied in the study of complementarity problems and variational inequalities in [Bulavski et al., 1998, 2001; Isac, 1999a, 2000b,c,d, 2001; Isac and Carbone, 1999; Isac and Kalashnikov, 2001;

Isac and Obuchowska, 1998; Isac et al., 1997, 2002; Kalashnikov and Isac, 2002; Zhao, 1998; Zhao and Han, 1999; Zhao and Isac, 2000b]. We note that the notion of an exceptional family of elements is based on the Leray–Schauder type alternatives.

This work can be considered as a starting point of a new research direction in the study of multivalued complementarity problems.

3.8.1   Preliminaries

Let $(H, \langle \cdot, \cdot \rangle)$ be a Hilbert space and $K \subset H$ a closed pointed convex cone.

Because K is closed and convex, the projection operator $P_K : H \to K$ onto K is well defined by the equation

$$\|x - P_K(x)\| = \min_{y \in K} \|x - y\|.$$

It is known that for every $x \in H$, $P_K(x)$ is uniquely defined by the relations:

(i) $\langle P_K(x) - x, y \rangle \geq 0$ for all $y \in K$.

(ii) $\langle P_K(x) - x, P_K(x) \rangle = 0$.

All topological vector spaces in this section are assumed to be real Hausdorff spaces. Let E, F be topological vector spaces, $X \subset E$ and $Y \subset F$. Denote by ∂X, $\overline{X}$, and $co(X)$ the boundary, the closure, and the convex hull of X, respectively, and by $\mathcal{P}(X)$ the family of all nonempty subsets of X.

Let $f : X \to Y$ be a set-valued mapping, that is, $f : X \to \mathcal{P}(Y)$. The mapping f is called *upper semicontinuous* (u.s.c.) on X if the set $\{x \in X \mid f(x) \subset V\}$ is open in X whenever V is an open subset of Y. f is said to be *compact* if $f(X)$ is relatively compact in Y.

A subset D of E is called *contractible* if there is a continuous mapping $h : D \times [0, 1] \to D$ with $h(x, 0) = x$ and $h(x, 1) = x_0$, for some $x_0 \in D$.

We note that if D is convex, it is contractible because for any $x_0 \in D$ we can consider $h(x, t) = t x_0 + (1 - t)x$. Similarly, a star-shaped set at x_0 is contractible to x_0. If $M \subset X$ is a nonempty subset, we say that a continuous mapping $r : X \to M$ is a *retraction* if and only if $r(x) = x$ for all $x \in M$. In this case we say that M is a *retract* of X. A set $D \subset X$ is called a *neighbourhood retract* if and only if D is a retract of some of its neighbourhoods.

A compact metric space M is called an *absolute neighbourhood retract* (ANR) if it has the universal property that every homeomorphic image of M in a separable metric space is a neighbourhood retract. Every compact convex set in an Euclidean space is an absolute neighbourhood retract. It is well known that if $f : X \to Y$ is u.s.c and $f(x)$ is compact for every $x \in K$, then for every compact subset D of X the set

$$f(D) = \bigcup_{x \in D} f(x)$$

is also compact [Berge, 1963].

If Ω is a lattice with a minimal element denoted by 0, a function $\Phi : \mathcal{P}(E) \rightarrow \Omega$ is called a *measure of noncompactness* provided that the following conditions hold for any $X_1, X_2 \in \mathcal{P}(E)$.

1. $\Phi(\overline{co}(X_1)) = \Phi(X_1)$.

2. $\Phi(X_1) = 0$ if and only if X_1 is precompact.

3. $\Phi(X_1 \cup X_2) = \max\{\Phi(X_1), \Phi(X_2)\}$.

We say that a mapping $f : X \rightarrow Y$ is Φ-condensing if for all $A \subset X$ with $\Phi(f(A)) \geq \Phi(A)$, A is relatively compact. A compact set-valued mapping $f : X \rightarrow E$ is Φ-condensing if either the domain X is complete or E is quasi-complete. Every set-valued mapping defined on a compact set is Φ-condensing (see [Ben-El-Mechaiekh and Idzik, 1998; Chebbi and Florenzano, 1995], and [Fitzpatrick and Petryshin, 1974]).

3.8.2 Approachable and Approximable Mappings

Let $E(\tau)$ be a Hausdorff locally convex topological vector space, $\mathfrak{U}$ be a fundamental basis of convex adjoint neighborhoods of the origin, and X, Y nonempty subsets of E. In this section we suppose that $f : X \rightarrow Y$ is a set-valued mapping with nonempty values.

We say that a single-valued mapping $s : X \rightarrow Y$ is a *selection* of the set-valued mapping $f : X \rightarrow Y$ if for any $x \in X$, $s(x) \in f(x)$.

In [Ben-El-Mechaiekh and Deguire, 1992; Ben-El-Mechaiekh and Idzik, 1998; Ben-El-Mechaiekh and Isac, 1998; Ben-El-Mechaiekh et al., 1994; Cellina, 1969], and [Gorniewicz et al., 1989] were introduced and studied the following notions.

For given $U, V \in \mathfrak{U}$, a function $s : X \rightarrow Y$ is called a (U, V)-*approximable selection* of f if for any $x \in X$, $s(x) \in (f[(x + U) \cap X] + V) \cap Y$.

The set-valued mapping $f : X \rightarrow Y$ is said to be *approachable* if it has a continuous (U, V)-approximable selection for any $(U, V) \in \mathfrak{U} \times \mathfrak{U}$.

Finally, we say that $f : X \rightarrow Y$ is *approximable* if its restriction $f|_D$ to any compact subset D of X is approachable. Examples of approachable and approximable mappings can be found in the above-mentioned references. Now we indicate a few examples.

If X is a topological space, Y a convex subset in a locally convex space, and $f : X \rightarrow Y$ an u.s.c. with convex values, then f is approximable. If X is a compact ANR, Y is an ANR, and the values of f are compact, then f is approachable.

3.8.3   Complementarity Problem

Let $(H, \langle \cdot, \cdot \rangle)$ be a Hilbert space, $K \subset H$ a closed pointed convex cone and $f : H \to H$ a set-valued mapping with nonempty values.

We say that f is *completely upper semicontinuous* (c.u.s.c.) if it is upper semicontinuous and for any bounded set $B \subset H$ we have that $f(B) = \bigcup_{x \in B} f(x)$ is relatively compact.

We say that f is *projectionally Φ-condensing (projectionally approximable)* with respect to K if $P_K(f)$ is Φ-condensing (resp., approximable).

The *multivalued complementarity problem* defined by f and the cone K is

$$MCP(f, K) : \begin{cases} \text{find } x_* \in K \quad \text{and} \\ x_*^f \in f(x_*) \cap K^* \quad \text{such that} \\ \langle x_*, x_*^f \rangle = 0. \end{cases}$$

3.8.4   Inversions of Set-Valued Mappings

Let $(H, \langle \cdot, \cdot \rangle)$ be a Hilbert space and $\| \cdot \|$ the norm generated by $\langle \cdot, \cdot \rangle$. We recall the following definition which is an extension of Example 5.1, p. 169 of [do Carmo, 1992]:

DEFINITION 3.87 *The operator*

$$i : H \backslash \{0\} \to H \backslash \{0\}; \; i(x) = \frac{x}{\|x\|^2}$$

is called the inversion *(of pole 0).*

It is easy to see that i is one to one and $i^{-1} = i$.

Let $K \subset H$ be a closed pointed convex cone and $f : K \to H$ a set-valued mapping. Because $K \backslash \{0\}$ is an invariant set of i the following definition makes sense.

DEFINITION 3.88 *The* inversion *(of pole 0) of the set-valued mapping f is the mapping $\mathcal{I}(f) : K \to H$ defined by:*

$$\mathcal{I}(f)(x) = \begin{cases} \|x\|^2 (f \circ i)(x) & \text{if} \quad x \neq 0, \\ \{0\} & \text{if} \quad x = 0, \end{cases}$$

where $\|x\|^2$ multiplies each element of $(f \circ i)(x)$.

We can show that $\mathcal{I}(\mathcal{I}(f)) = f$.

Let $C \subseteq H$ be a set which contains at least one nonisolated point, $F, G : C \to H$ be set-valued mappings, and x_0 a nonisolated point of C. The following definition is an extension of Definition 2.2 of [Nemeth, 1992].

DEFINITION 3.89 *The limit*

$$\underline{F}^{\#}(x_0) = \liminf_{\substack{x \to x_0, x \in C \\ x^F \in F(x), x_0^F \in F(x_0)}} \frac{\langle x^F - x_0^F, x - x_0 \rangle}{\|x - x_0\|^2}$$

is called the lower scalar derivative *of F at x_0. Taking* lim sup *in place of* lim inf, *we can define the upper scalar derivative* $\overline{F}^{\#}(x_0)$ *of F at x_0 similarly.*

Definition 3.89 can be extended for the unordered pair of set-valued mappings (F, G). The idea was inspired by the notion of derivative of a function with respect to another function [Choquet, 1969].

DEFINITION 3.90 *The limit*

$$\underline{(F, G)}^{\#}(x_0) = \liminf_{\substack{x \to x_0, x \in C \\ x^F \in F(x), x_0^F \in F(x_0) \\ x^G \in G(x), x_0^G \in G(x_0)}} \frac{\langle x^F - x_0^F, x^G - x_0^G \rangle}{\|x - x_0\|^2}$$

is called the lower scalar derivative *of the unordered pair of set-valued mappings (F, G) at x_0. Taking* lim sup *in place of* lim inf, *we can define the upper scalar derivative* $\overline{(F, G)}^{\#}(x_0)$ *of (F, G) at x_0 similarly.*

REMARK 3.7 *If $G = I$, we obtain* Definition 3.89.

3.8.5 Exceptional Family of Elements

The next definition can be found in [Isac, 1999b].

DEFINITION 3.91 *Let $(H, \langle \cdot, \cdot \rangle)$ be a Hilbert space, $K \subset H$ be a closed pointed convex cone, and $f : H \to H$ a set-valued mapping. We say that a family of elements $\{x_r\}_{r>0} \subset K$ is an* exceptional family of elements *for f with respect to K, if for every real number $r > 0$, there exists a real number $\mu_r > 0$ and an element $x_r^f \in f(x_r)$ such that the following conditions are satisfied.*

1. $u_r = \mu_r x_r + x_r^f \in K^*$ *for all $r > 0$.*

2. $\langle u_r, x_r \rangle = 0$ *for all $r > 0$.*

3. $\|x_r\| \to +\infty$ *as $r \to +\infty$.*

The next theorem is Theorem 2 of [Isac, 1999b].

THEOREM 3.92 *Let $(H, \langle \cdot, \cdot \rangle)$ be a Hilbert space, $K \subset H$ a closed pointed convex cone, and $f : H \to H$ an u.s.c set-valued mapping with nonempty values. If the following assumptions are satisfied,*

1. $x - f(x)$ is projectionally Φ-condensing, or $f(x) = x - T(x)$, where T is a c.u.s.c. set-valued mapping with nonempty values;

2. $x - f(x)$ is projectionally approximable and $P_K[x - f(x)]$ has closed values;

then there exists either a solution to the problem $MCP(f, K)$, or an exceptional family of elements for f with respect to K.

The next definition can be found in [Isac, 1999b].

DEFINITION 3.93 *Let $(H, \langle \cdot, \cdot \rangle)$ be a Hilbert space, and $K \subset H$ a closed pointed convex cone. We say that a set-valued mapping $f : H \to H$ with nonempty values satisfies* condition Θ *with respect to K if there exists a real number $\rho > 0$ such that for each $x \in K$ with $\|x\| > \rho$ there exists $p \in K$ with $\|p\| < \|x\|$ such that $\langle x - p, x^f \rangle \geq 0$ for all $x^f \in f(x)$.*

DEFINITION 3.94 *Let $(H, \langle \cdot, \cdot \rangle)$ be a Hilbert space, and $K \subset H$ a closed pointed convex cone. We say that a set-valued mapping $f : H \to H$ with nonempty values satisfies* condition $\tilde{\Theta}$ *with respect to K if there exists a real number $\rho > 0$ such that for each $x \in K$ there exists $p \in K$ with $\langle p, x \rangle < \|x\|^2$ such that $\langle x - p, x^f \rangle \geq 0$ for all $x^f \in f(x)$.*

The next lemma shows that condition $\tilde{\Theta}$ is an extension of condition Θ.

LEMMA 3.95 *Let $(H, \langle \cdot, \cdot \rangle)$ be a Hilbert space, $K \subset H$ a closed convex cone, and $f : H \to H$ a set-valued mapping with nonempty values. If f satisfies condition Θ with respect to K, then it satisfies the condition $\tilde{\Theta}$ with respect to K.*

Proof. Because f satisfies condition Θ with respect to K, there exists $\rho > 0$ such that for each $x \in K$ with $\|x\| > \rho$, there exists $p \in K$ with $\|p\| < \|x\|$ such that $\langle x - p, x^f \rangle \geq 0$ for all $x^f \in f(x)$. By the Cauchy inequality

$$\langle p, x \rangle \leq \|p\| \|x\| < \|x\|^2.$$

Hence, f satisfies condition $\tilde{\Theta}$ with respect to K. $\square$

The next theorem is proved in [Isac, 1999b].

THEOREM 3.96 *Let H be a Hilbert space, $K \subset H$ a closed pointed convex cone, and $f : H \to H$ a set-valued mapping with nonempty values. If f satisfies condition Θ with respect to K, then it is without an exceptional family of elements with respect to K.*

THEOREM 3.97 *Let H be a Hilbert space, $K \subset H$ a closed pointed convex cone, and $f : H \to H$ a set-valued mapping with nonempty values. If f*

satisfies condition $\tilde{\Theta}$ with respect to K, then it is without an exceptional family of elements with respect to K.

Proof. Suppose to the contrary, that f has an exceptional family of elements $\{x_r\}_{r>0} \subset K$ with respect to K. Because $\|x_r\| \to \infty$ as $r \to \infty$, we can choose a real number r such that $\|x_r\| > \rho$. By condition $\tilde{\Theta}$ there exists $p_r \in K$ such that $\langle p_r, x_r \rangle < \|x_r\|^2$ and

$$\langle x_r - p_r, x^f \rangle \geq 0, \quad \text{for all } x^f \in f(x_r). \tag{3.52}$$

By the definition of the exceptional family of elements there exists $\mu_r > 0$ and $x_r^f \in f(x_r)$ such that

$$\begin{cases} u_r = \mu_r x_r + x_r^f \in K^* \\ \text{and} \\ \langle u_r, x_r \rangle = 0. \end{cases} \tag{3.53}$$

By Equations (3.52) and (3.53) we have

$$0 \leq \langle x_r - p_r, x_r^f \rangle = \langle x_r - p_r, u_r - \mu_r x_r \rangle \tag{3.54}$$

$$= \langle x_r - p_r, u_r \rangle - \mu_r \|x_r\|^2 + \mu_r \langle p_r, x_r \rangle \leq -\mu_r (\|x_r\|^2 - \langle p_r, x_r \rangle) < 0, \tag{3.55}$$

which is a contradiction.

3.8.6   Infinitesimal Exceptional Family of Elements

DEFINITION 3.98 *Let $(H, \langle \cdot, \cdot \rangle)$ be a Hilbert space, $K \subset H$ a closed pointed convex cone, and $g : K \to H$ a set-valued mapping with nonempty values. We say that $\{y_r\}_{r>0} \subset K$ is an infinitesimal exceptional family of elements for g with respect to K, if for every real number $r > 0$, there exists a real number $\mu_r > 0$ and an element $y_r^g \in g(y_r)$ such that the following conditions are satisfied.*

1. $v_r = \mu_r y_r + y_r^g \in K^*$.

2. $\langle v_r, y_r \rangle = 0$.

3. $y_r \to 0$ as $r \to +\infty$.

DEFINITION 3.99 *Let $(H, \langle \cdot, \cdot \rangle)$ be a Hilbert space, and $K \subset H$ a closed pointed convex cone. We say that a set-valued mapping $g : H \to H$ with nonempty values satisfies condition ${}^i\Theta$ with respect to K if there exists a real number $\lambda > 0$ such that for each $y \in K \backslash \{0\}$ with $\|y\| < \lambda$, there exists $q \in K$ with $\|q\| < \|y\|$ such that*

$$\langle y - q, y^g \rangle \geq 0, \tag{3.56}$$

for all $y^g \in g(y)$.

DEFINITION 3.100 *Let* $(H, \langle \cdot, \cdot \rangle)$ *be a Hilbert space, and* $K \subset H$ *a closed pointed convex cone. We say that a set-valued mapping* $g : H \to H$ *with nonempty values satisfies* condition $^i\tilde{\Theta}$ *with respect to* K *if there exists a real number* $\lambda > 0$ *such that for each* $y \in K \backslash \{0\}$ *with* $\|y\| < \lambda$, *there exists* $q \in K$ *with* $\langle q, y \rangle < \|y\|^2$ *such that*

$$\langle y - q, y^g \rangle \geq 0, \tag{3.57}$$

for all $y^g \in g(y)$.

The next lemma shows that condition $^i\tilde{\Theta}$ is an extension of condition $^i\Theta$ and it can be proved similarly to Lemma 3.95.

LEMMA 3.101 *Let* $(H, \langle \cdot, \cdot \rangle)$ *be a Hilbert space,* $K \subset H$ *a closed pointed convex cone, and* $g : H \to H$ *a set-valued mapping with nonempty values. If* g *satisfies condition* $^i\Theta$ *with respect to* K, *then it satisfies condition* $^i\tilde{\Theta}$ *with respect to* K.

THEOREM 3.102 *Let* $(H, \langle \cdot, \cdot \rangle)$ *be a Hilbert space,* $K \subset H$ *a closed pointed convex cone, and* $g : H \to H$ *a set-valued mapping with nonempty values. If* g *satisfies condition* $^i\tilde{\Theta}$ *with respect to* K, *then it is without an infinitesimal exceptional family of elements with respect to* K.

Proof. Suppose to the contrary, that g has an infinitesimal family of elements $\{y_r\}_{r>0} \subset K$ with respect to K. For any $r > 0$ such that $\|y_r\| < \rho$ there is an element $q_r \in K$ with $\langle q_r, y_r \rangle < \|y_r\|^2$ satisfying relation (3.56); that is,

$$\langle y_r - q_r, y_r^g) \rangle \geq 0.$$

for an arbitrary $y_r^g \in g(y_r)$. Because, according to Definition 3.98, $\langle v_r, y_r \rangle = 0$ and $v_r \in K^*$, we have

$$
\begin{aligned}
0 \leq \langle y_r - q_r, y_r^g \rangle &= \langle y_r - q_r, v_r - \mu_r y_r \rangle \\
&= -\mu_r \|y_r\|^2 - \langle q_r, v_r \rangle + \mu_r \langle q_r, y_r \rangle \\
&\leq -\mu_r (\|y_r\|^2 - \langle q_r, y_r \rangle) < 0,
\end{aligned}
$$

which is a contradiction.

REMARK 3.8 *At first sight Theorem 3.102 seems to be a direct consequence of Theorems 3.105 and 3.107, proved in the next section. However, note that there might be an infinitesimal family of elements of* g *which contains zero.*

COROLLARY 3.103 *Let* $(H, \langle \cdot, \cdot \rangle)$ *be a Hilbert space,* $K \subset H$ *a closed pointed convex cone, and* $g : H \to H$ *a set-valued mapping with nonempty* *values. If g satisfies condition $^i\Theta$ with respect to K, then it is without an infinitesimal exceptional family of elements with respect to K.*

Proof. By Lemma 3.101 g satisfies condition $^i\tilde{\Theta}$ with respect to K. Hence, by Theorem 3.102 g is without an infinitesimal exceptional family of elements with respect to K. ∎

REMARK 3.9 *At first sight Corollary 3.103 seems to be a direct consequence of Theorems 3.105 and 3.108, proved in the next section.. However, note that there might be an infinitesimal family of elements of g which contains zero.*

3.8.7   A Duality and Main results

THEOREM 3.104 *Let $(H, \langle \cdot, \cdot \rangle)$ be a Hilbert space, $K \subset H$ a closed convex cone, and $f : K \to H$ a mapping. Then $(x_*, x_*^f) \notin \{0\} \times H$ is a solution of $MCP(f, K)$ if and only if (y_*, y_*^g) is a solution of $MCP(g, K)$, where $y_* = i(x_*)$ is the inversion of x_*,*

$$y_*^g = \frac{1}{\|x_*\|^2} x_*^f,$$

and $g = \mathcal{I}(f)$ is the inversion of f.

Proof. First we have to prove that $y_*^g \in g(y_*)$. Dividing both sides of the relation $x_*^f \in f(x_*)$ by $\|x_*\|^2$ we obtain

$$y_*^g \in \frac{1}{\|x_*\|^2} f(x_*),$$

which implies $y_*^g \in \|y_*\|^2 f(i(y_*)) = \mathcal{I}(f)(y_*) = g(y_*)$. It is easy to see that

$$\langle y_*, y_*^g \rangle = \frac{1}{\|x_*\|^4} \langle x_*, x_*^f \rangle \tag{3.58}$$

and

$$\langle y_*^g, z \rangle = \frac{1}{\|x_*\|^2} \langle x_*^f, z \rangle, \tag{3.59}$$

for every $z \in K$. By using (3.58),

$$\langle x_*, x_*^f \rangle = 0$$

if and only if

$$\langle y_*, y_*^g \rangle = 0.$$

By using (3.59), $x_*^f \in K^*$ if and only if $y_*^g \in K^*$.

THEOREM 3.105 *Let* $(H, \langle \cdot, \cdot \rangle)$ *be a Hilbert space,* $K \subset H$ *a closed pointed convex cone, and* $f : K \to H$ *a set-valued mapping with nonempty values.* $\{x_r\}_{r>0} \subset K \backslash \{0\}$ *is an exceptional family of elements for* f *with respect to* K *if and only if* $\{y_r\}_{r>0} \subset K \backslash \{0\}$ *is an infinitesimal exceptional family of elements for* g *with respect to* K*, where* $y_r = i(x_r)$ *and* $g = \mathcal{I}(f)$*.* $\square$

Proof. Bearing in mind the notations of Definition 3.98, we have

$$v_r = \mu_r y_r + y_r^g,$$

for some $y_r^g \in g(y_r)$. Hence,

$$v_r = \|y_r\|^2 \left(\mu_r i(y_r) + \frac{y_r^g}{\|y_r\|^2} \right).$$

Because $i^{-1} = i$, we have

$$v_r = \frac{1}{\|x_r\|^2} \left(\mu_r x_r + \|x_r\|^2 y_r^g \right). \tag{3.60}$$

Let

$$x_r^f := \|x_r\|^2 y_r^g. \tag{3.61}$$

We have $x_r^f \in f(x_r)$. Indeed,

$$x_r^f \in \|x_r\|^2 g(y_r) = \|x_r\|^2 \mathcal{I}(f)(y_r) = \|x_r\|^2 \|y_r\|^2 f(i(y_r)) = f(x_r).$$

Now let

$$u_r = \mu_r x_r + x_r^f. \tag{3.62}$$

Equations (3.60) through (3.62) imply that

$$v_r = \frac{1}{\|x_r\|^2} u_r.$$

Therefore,

$$\langle v_r, y_r \rangle = \frac{1}{\|x_r\|^4} \langle u_r, x_r \rangle \tag{3.63}$$

and

$$\langle v_r, z \rangle = \frac{1}{\|x_r\|^2} \langle u_r, z \rangle, \tag{3.64}$$

for every $z \in K$. Because $\|x_r\| \cdot \|y_r\| = 1$, $\|x_r\| \to +\infty$ if and only if $y_r \to 0$. By using (3.63),

$$\langle u_r, x_r \rangle = 0$$

if and only if

$$\langle v_r, y_r \rangle = 0.$$

By using (3.64), $u_r \in K^*$ if and only if $v_r \in K^*$.　　　　　□

THEOREM 3.106 *Let* $(H, \langle \cdot, \cdot \rangle)$ *be a Hilbert space,* $K \subset H$ *a closed pointed convex cone, and* $f : H \to H$ *an u.s.c set-valued mapping with nonempty values such that*

1. $x - f(x)$ *is projectionally* Φ-*condensing, or* $f(x) = x - T(x)$, *where* T *is a c.u.s.c. set-valued mapping with nonempty values.*

2. $x - f(x)$ *is projectionally approximable and* $P_K[x - f(x)]$ *is with closed values.*

3. $f(0) \cap K^* = \emptyset$.

If every infinitesimal exceptional family of elements for $g = \mathcal{I}(f)$ *with respect to* K *contains* 0, *then the multivalued complementarity problem* $MCP(f, K)$ *has a nonzero solution.*

Proof. Because $f(0) \cap K^* = \emptyset$, if $MCP(f, K)$ has a solution, then this solution is nonzero. By Theorem 3.92, it is enough to prove that f is without an exceptional family of elements with respect to K. Suppose to the contrary that $\{x_r\}_{r>0}$ is an exceptional family of elements for f with respect to K. Because $f(0) \cap K^* = \emptyset$, by the definition of an exceptional family of elements $\{x_r\}_{r>0} \subset K \backslash \{0\}$. Hence, by Theorem 3.105, $g = \mathcal{I}(f)$ has an infinitesimal exceptional family of elements with respect to K which does not contain 0, which is a contradiction.　　　　　□

THEOREM 3.107 *Let* $(H, \langle \cdot, \cdot \rangle)$ *be a Hilbert space,* $K \subset H$ *a closed pointed convex cone,* $f : H \to H$ *a set-valued mapping with nonempty values, and* $g = \mathcal{I}(f)$. *Then,* f *satisfies condition* $\tilde{\Theta}$ *with respect to* K *if and only if* g *satisfies the condition* $^i\tilde{\Theta}$ *with respect to* K.

Proof. Suppose that g satisfies condition $^i\tilde{\Theta}$ with respect to K and prove that f satisfies condition $\tilde{\Theta}$ with respect to K. Consider the constant λ of condition $^i\tilde{\Theta}$ and let

$$\rho = \frac{1}{\lambda}.$$

Let $x \in K$ with

$$\|x\| > \rho \tag{3.65}$$

and $y = i(x)$. Because

$$\|y\| = \frac{1}{\|x\|},$$

it follows that $\|y\| < \lambda$. Hence, by condition ${}^i\tilde{\Theta}$, there exists $q \in K$ with $\langle q, y \rangle < \|y\|^2$ such that $\langle y - q, y^g \rangle \geq 0$ for all $y^g \in g(y)$. Let

$$p = \frac{q}{\|y\|^2}. \tag{3.66}$$

Because $\langle q, y \rangle < \|y\|^2$ and $i^{-1} = i$, relation (3.66) implies that

$$\langle p, x \rangle = \frac{\langle q, y \rangle}{\|y\|^4} < \frac{1}{\|y\|^2} = \|x\|^2. \tag{3.67}$$

Now let $x_f := \|x\|^2 y^g$. We have $x_r^f \in f(x_r)$. Indeed,

$$x_r^f \in \|x_r\|^2 g(y_r) = \|x_r\|^2 \mathcal{I}(f)(y_r) = \|x_r\|^2 \|y_r\|^2 f(i(y_r)) = f(x_r).$$

Hence,

$$\langle x - p, x^f \rangle = \|x\|^2 \langle x - p, y^g \rangle = \|x\|^4 \langle y - q, y^g \rangle \geq 0. \tag{3.68}$$

By (3.65), (3.67), and (3.68) f satisfies condition $\tilde{\Theta}$ with respect to K. Now, suppose that f satisfies condition $\tilde{\Theta}$ with respect to K and prove that g satisfies condition ${}^i\tilde{\Theta}$ with respect to K. Consider the constant $\rho > 0$ of condition $\tilde{\Theta}$ and let

$$\lambda = \frac{1}{\rho}.$$

Let $y \in K \setminus \{0\}$ with $\|y\| < \lambda$.

We have to prove that there exists $q \in K$ with $\langle q, y \rangle < \|y\|^2$ such that $\langle y - q, y^g \rangle \geq 0$, for all $y^g \in g(y)$. Since $f = \mathcal{I}(g)$, we can proceed as above. $\square$

The next theorem can be proved similarly to Theorem 3.107.

THEOREM 3.108 *Let* $(H, \langle \cdot, \cdot \rangle)$ *be a Hilbert space,* $K \subset H$ *a closed pointed convex cone,* $f : H \to H$ *a set-valued mapping with nonempty values, and* $g = \mathcal{I}(f)$. *Then,* f *satisfies condition* Θ *with respect to* K *if and only if* g *satisfies condition* ${}^i\Theta$ *with respect to* K.

THEOREM 3.109 *Let* H *be a Hilbert space,* $K \subset H$ *a closed pointed convex cone, and* $f : H \to H$ *an u.s.c set-valued mapping with nonempty values such that*

1. $x - f(x)$ *is projectionally* Φ-*condensing, or* $f(x) = x - T(x)$, *where* T *is a c.u.s.c. set-valued mapping with nonempty values.*

2. $x - f(x)$ *is projectionally approximable and* $P_K[x - f(x)]$ *has closed values.*

If $g = \mathcal{I}(f)$ satisfies condition $^i\Theta$ with respect to K, then the multivalued complementarity problem $MCP(f, K)$ has a solution.

Proof. By Theorem 3.108, f satisfies condition Θ with respect to K. Hence, Theorems 3.96 and 3.92 imply that the multivalued complementarity problem $MCP(f, K)$ has a solution. $\qquad\square$

THEOREM 3.110 *Let H be a Hilbert space, $K \subset H$ a closed pointed convex cone, and $f : H \to H$ an u.s.c set-valued mapping with nonempty values such that*

1. $x - f(x)$ *is projectionally Φ-condensing, or $f(x) = x - T(x)$, where T is a c.u.s.c. set-valued mapping with nonempty values.*

2. $x - f(x)$ *is projectionally approximable and $P_K[x - f(x)]$ has closed values.*

If $g = \mathcal{I}(f)$ satisfies condition $^i\tilde{\Theta}$ with respect to K, then the multivalued complementarity problem $MCP(f, K)$ has a solution.

Proof. By Theorem 3.107, f satisfies condition $\tilde{\Theta}$ with respect to K. Hence, Theorem 3.97 implies that the multivalued complementarity problem $MCP(f, K)$ has a solution. $\qquad\square$

THEOREM 3.111 *Let $(H, \langle \cdot, \cdot \rangle)$ be a Hilbert space, $K \subset H$ a closed convex cone, and $f : H \to H$ an u.s.c set-valued mapping with nonempty values such that*

1. $x - f(x)$ *is projectionally Φ-condensing, or $f(x) = x - T(x)$, where T is a c.u.s.c. set-valued mapping with nonempty values.*

2. $x - f(x)$ *is projectionally approximable and $P_K[x - f(x)]$ has closed values.*

If there is a $\delta > 0$ and a mapping $h : B(0, \delta) \cap K \to H$ with $h(0) = 0$ and

$$\begin{cases} \overline{h}^{\#}(0) < 1, \\ \underline{(I - h, \mathcal{I}(f))}^{\#}(0) > 0, \end{cases}$$

where $B(0, \delta) = \{z \in H : \|z\| < \delta\}$, then the multivalued complementarity problem $MCP(f, K)$ has a solution.

Proof. Let $g = \mathcal{I}(f)$. Because $\overline{h}^{\#}(0) < 1$, there is a λ_1 with $0 < \lambda_1 < \delta$ such that for every $y \in K$ with $\|y\| < \lambda_1$ we have

$$\langle h(y), y \rangle < \|y\|^2. \tag{3.69}$$

Inasmuch as

$$(I - h, g)^{\#}(0) > 0,$$

there is a λ_2 with $0 < \lambda_2 < \delta$ such that for every $y \in K$ with $\|y\| < \lambda_2$ we have

$$\langle y - h(y), y^g \rangle > 0, \tag{3.70}$$

for all $y^g \in g(y)$. Let $\lambda = \min\{\lambda_1, \lambda_2\}$. Obviously,

$$\lambda > 0. \tag{3.71}$$

For

$$\|y\| < \lambda \tag{3.72}$$

let $q = h(y)$. Then, relations (3.69) and (3.70) imply

$$\langle q, y \rangle < \|y\|^2. \tag{3.73}$$

and

$$\langle y - q, y^g \rangle \geq 0, \tag{3.74}$$

respectively, for all $y^g \in g(y)$. Hence, relations (3.71) through (3.74) imply that g satisfies condition $^i\Theta$. Hence, Theorem 3.110 implies that the problem NCP(f, K) has a solution.   □

In the particular case $h = 0$ we have as follows.

COROLLARY 3.112 *Let* $(H, \langle \cdot, \cdot \rangle)$ *be a Hilbert space,* $K \subset H$ *a closed pointed convex cone, and* $f : H \to H$ *an u.s.c set-valued mapping with nonempty values such that*

1. $x - f(x)$ *is projectionally* Φ-*condensing, or* $f(x) = x - T(x)$, *where* T *is a c.u.s.c. set-valued mapping with nonempty values.*

2. $x - f(x)$ *is projectionally approximable and* $P_K[x - f(x)]$ *has closed values.*

If $\mathcal{I}(f)^{\#}(0) > 0$, *then the multivalued complementarity problem* $MCP(f, K)$ *has a solution.*

3.9   The Asymptotic Browder–Hartman–Stampacchia Condition and Interior Bands of ε-Solutions for Nonlinear Complementarity Problems

Let $(\mathbb{R}^n, \langle \cdot, \cdot \rangle)$ be the n-dimensional Euclidean space ordered by the closed pointed convex cone $\mathbb{R}^n_+$ and $f : \mathbb{R}^n \to \mathbb{R}^n$ a continuous function. We consider the following nonlinear complementarity problem.

$$\text{NCP}(f, \mathbb{R}^n_+) : \begin{cases} \text{find } x^0 \in \mathbb{R}^n_+ \text{ such that} \\ f(x^0) \in \mathbb{R}^n_+ \text{ and } \langle x^0, f(x^0) \rangle = 0. \end{cases}$$

It is well known that the $\mathrm{NCP}(f, \mathbb{R}^n_+)$ has many applications in optimization, economics, engineering, game theory and mechanics [Cottle et al., 1992; Ferris and Pang, 1997; Harker and Pang, 1990; Isac, 1992, 2000d].

We note that there exist several equivalent formulations of the $\mathrm{NCP}(f, \mathbb{R}^n_+)$. In particular, several formulations are in the form of a nonlinear equation of the form $F(x) = 0$, where $F : \mathbb{R}^n \to \mathbb{R}^n$ is a continuous function. By using such formulations, several techniques proposed by some authors are based on the idea to perturb F to a certain $F(x, \varepsilon)$, where ε is a positive parameter and to consider the equation $F(x, \varepsilon) = 0$. If $F(x, \varepsilon) = 0$ has a unique solution, denoted by $x(\varepsilon)$ and $x(\varepsilon)$ is continuous in ε, then the solutions describe (depending on the properties of F) a short path denoted $\{x(\varepsilon) : \varepsilon \in]0, \varepsilon_0]\}$ or a long path $\{x(\varepsilon) : \varepsilon \in]0, \infty[\}$. We note that, if a short path $\{x(\varepsilon) : \varepsilon \in]0, \varepsilon_0]\}$ is bounded, then for any sequence $\{\varepsilon_k\}$ with $\{\varepsilon_k\} \to 0$, the sequence $\{x(\varepsilon_k)\}$ has at least one accumulation point, which by continuity is a solution to the $\mathrm{NCP}(f, \mathbb{R}^n_+)$.

Based on this fact, several numerical methods for solving the $\mathrm{NCP}(f, \mathbb{R}^n_+)$ have been developed, for example, the interior-point path-following methods, regularization methods, and noninterior path-following methods among others. About such methods the reader can see the papers [Burke and Xu, 1998, 2000; Chen and Chen, 1999; Chen and Mangasarian, 1996; Chen et al., 1997; Facchini, 1998; Facchini and Kanzow, 1999; Ferris and Pang, 1997; Gowda and Tawhid, 1999; Guler, 1990, 1993; Harker and Pang, 1990; Hotta and Yoshise, 1999; Kanzow, 1996; Kojima et al., 1991a,b; Megiddo, 1989; Monteiro and Adler, 1989], and [Tseng, 1997]. The most common interior-point path-following method is based on the notion of the *central path*. We recall [Zhao and Isac, 2000a] that the curve $\{x(\varepsilon) : \varepsilon \in]0, \infty[\}$ is said to be the central path if for each $\varepsilon > 0$ the vector $x(\varepsilon)$ is the unique solution to the system

$$\begin{cases} x(\varepsilon) > 0, \; f(x(\varepsilon)) > 0 \\ \text{and } X(\varepsilon)f(x(\varepsilon)) = \varepsilon e, \end{cases} \tag{3.75}$$

where the inequality $>$ means that the components of the vector are strictly positive, $e = (1, \ldots, 1)^T$, $X(\varepsilon) = $ the matrix $\mathrm{diag}(x(\varepsilon))$, and $x(\varepsilon)$ is continuous on $]0, \infty[$. It is well known that, for a general $\mathrm{NCP}(f, \mathbb{R}^n_+)$, the system (3.74) may have multiple solutions for a given $\varepsilon > 0$, and even if the solution is unique it is not necessarily continuous in ε. Therefore, the existence of the central path is not always guaranteed.

We consider in this section the multivalued mapping $\mathcal{U} :]0, \infty[\to \mathcal{S}(\mathbb{R}^n_{++})$ defined by

$$\mathcal{U}(\varepsilon) = \{x \in \mathbb{R}^n_{++} : f(x) > 0, Xf(x) = \varepsilon e\},$$

where $X = $ the matrix $\mathrm{diag}(x)$, $\mathcal{S}(\mathbb{R}^n_{++})$ is the collection of all subsets of $\mathbb{R}^n_{++}$, and $\mathbb{R}^n_{++} = \{x = (x_1, \ldots, x_n) : x_1 > 0, \ldots, x_n > 0\}$. We say that $\mathcal{U}$ is the *interior band mapping*. The multivalued mapping $\mathcal{U}$ was studied from several points of view in [Zhao and Isac, 2000a].

Now, the main goal of our section is to study under what conditions the multivalued mapping $\mathcal{U}$ has the following properties:

(a) $\mathcal{U}(\varepsilon) \neq \emptyset$ for each $\varepsilon \in]0, \infty[$.

(b) For any fixed $\varepsilon_0 > 0$ the set $\bigcup_{\varepsilon \in]0,\varepsilon_0]} \mathcal{U}(\varepsilon)$ is bounded.

Conditions (a) and (b) were defined in [Zhao and Isac, 2000a].

Our study is based on the *asymptotic Browder–Hartman–Stampacchia condition*, on the notion of *scalar derivative* and the notion of *infinitesimal interior-point-ε-exceptional family of elements* for the *inversion* of f. By the results presented in this section we show another utility of the Browder–Hartman–Stampacchia condition, never put in evidence by other authors.

3.9.1   Preliminaries

For more details about the notions and the results presented in this section, the reader is referred to [Zhao and Isac, 2000a].

DEFINITION 3.113 *Let* $f : \mathbb{R}^n \to \mathbb{R}^n$ *be a continuous function. Given a scalar* $\varepsilon > 0$ *we say that a family* $\{x^r\}_{r>0} \subset \mathbb{R}^n_{++}$ *is an* interior-point-ε-exceptional family *for* f *if* $\|x^r\| \to \infty$ *as* $r \to \infty$ *and for each* x^r *there exists a positive number* $0 < \mu_r < 1$ *such that*

$$f_i(x^r) = \frac{1}{2}\left(\mu_r - \frac{1}{\mu_r}\right)x_i^r + \frac{\varepsilon\mu_r}{x_i^r}, \qquad (3.76)$$

for all $i = 1, 2, \ldots, n$.

THEOREM 3.114 *Let* $f : \mathbb{R}^n \to \mathbb{R}^n$ *be a continuous function. Then for each* $\varepsilon > 0$ *there exists either a point* $x(\varepsilon)$ *such that*

$$x(\varepsilon) > 0, \ f(x(\varepsilon)) > 0, \ x_i(\varepsilon)f_i(x(\varepsilon)) = \varepsilon, \ i = 1, 2, \ldots, n, \qquad (3.77)$$

or an interior-point-ε-exceptional family for f.

Proof. This result is proved in [Zhao and Isac, 2000a].   □

Let $K \subset \mathbb{R}^n$ be a closed pointed convex cone and $f : \mathbb{R}^n \to \mathbb{R}^n$ a continuous function. The following definition is Definition 3.50 where H is the n-dimensional Euclidean space.

DEFINITION 3.115 *We say that a family of elements* $\{x^r\}_{r>0} \subset K$ *is an exceptional family of elements (denoted EFE) for* f *with respect to* K *if* $\|x^r\| \to \infty$ *as* $r \to \infty$ *and for every real number* $r > 0$, *there exists a real number* $\mu_r > 0$ *such that the vector* $u^r = \mu_r x^r + f(x^r)$ *satisfies the following conditions.*

1. $u^r \in K^*$ *(the dual of K).*

2. $\langle u^r, x^r \rangle = 0.$

If in the definition of the problem $\mathrm{NCP}(f, \mathbb{R}^n_+)$ we replace the cone $\mathbb{R}^n_+$ by the cone K, we obtain the problem $\mathrm{NCP}(f, K)$, that is,

$$\mathrm{NCP}(f, K) : \begin{cases} \text{find } x^0 \in K \text{ such that} \\ f(x^0) \in K^* \text{ and } \langle x^0, f(x^0) \rangle = 0. \end{cases}$$

THEOREM 3.116 *If $f : \mathbb{R}^n \to \mathbb{R}^n$ is a continuous function and $K \subset \mathbb{R}^n$ is a closed pointed convex cone, then there exists either a solution to the $\mathrm{NCP}(f, K)$ or an exceptional family of elements for f with respect to K.*

Proof. A proof of this result is given in [Isac et al., 1997].   □

The following definition and theorem are exactly Definition 3.52 and Theorem 3.55.

DEFINITION 3.117 *Let $(H, \langle \cdot, \cdot \rangle)$ be a Hilbert space, $K \subset H$ a closed convex cone, and $f : H \to H$ a mapping. We say that the mapping f satisfies* condition Θ *with respect to K if*

$$\begin{cases} \text{there exists } \rho > 0 \text{ such that for each } x \in K \text{ with } \|x\| > \rho, \\ \text{there exists } p \in K \text{ with } \|p\| < \|x\| \text{ such that} \\ \langle x - p, f(x) \rangle \geq 0. \end{cases} \tag{3.78}$$

THEOREM 3.118 *Let H be a Hilbert space, $K \subset H$ a closed convex cone, and $f : H \to H$ a mapping. If f satisfies condition Θ with respect to K, then it is without an exceptional family of elements with respect to K.*

3.9.2   The Browder–Hartman–Stampacchia Condition

First we recall a general classical result. Let $(E, \| \cdot \|)$ be a reflexive Banach space and $f : E \to E^*$. We say that f is *hemicontinuous* if it is continuous from the line segment to the weak topology of E^*.

We say that f is *monotone* if for any $x, y \in E$ we have that

$$\langle x - y, f(x) - f(y) \rangle \geq 0,$$

and we say that f is *strongly monotone* if there exists a constant $\alpha > 0$ such that for any $x, y \in E$ we have

$$\langle x - y, f(x) - f(y) \rangle \geq \alpha \|x - y\|^2.$$

DEFINITION 3.119 *We say that f satisfies the* Browder–Hartman–Stampacchia condition *(denoted BHS) on a closed convex cone $K \subset E$ if there exists $\rho > 0$ such that $\langle x, f(x) \rangle > 0$, for any $x \in K$ with $\|x\| = \rho$.*

The following result is to support the importance of condition BHS.

THEOREM 3.120 (BROWDER–HARTMAN–STAMPACCHIA) *Let $(E, \| \cdot \|)$ be a reflexive Banach space, $f : E \to E^*$ a monotone hemicontinuous mapping, and $K \subset E$ a closed convex cone. If f satisfies condition BHS on K, then the problem $\mathrm{NCP}(f, K)$ has a solution.*

Proof. A proof of this result is given in [Isac, 2000d], Theorem 4.6. $\quad\square$

PROPOSITION 3.7 *If $f : E \to E^*$ is strongly monotone on K, then f satisfies condition BHS on K.*

Proof. Indeed, if we take $y = 0$, we have

$$\langle x, f(x) \rangle \geq \langle x, f(0) \rangle + \alpha \|x\|^2.$$

If $f(0) = 0$ we deduce that $\langle x, f(x) \rangle \geq \alpha \|x\|^2$, for any $x \in K$. If we take an arbitrary $\rho > 0$, we have that $\langle x, f(x) \rangle > 0$, for any $x \in K$ with $\|x\| = \rho$. If $f(0) \neq 0$, then in this case we consider the set

$$D = \left\{ x \in K : \|x\| \leq \frac{\|f(0)\|}{\alpha} \right\},$$

which is nonempty and bounded. For any $x \in K \backslash D$ we have

$$\alpha \|x\|^2 > \|x\| \cdot \|f(0)\| \geq -\langle x, f(0) \rangle,$$

which implies $\langle x, f(x) \rangle > 0$ for any $x \in K \backslash D$. Because D is bounded, there exists $\rho > 0$ such that $D \subset B(0, \rho)$ and for any $x \in K$ with $\|x\| = \rho$ we have $\langle x, f(x) \rangle > 0$. $\quad\square$

By the next result we can obtain many functions that satisfy condition BHS. Let $K \subset \mathbb{R}^n$ be a closed convex cone. We say that a function $T : K \to \mathbb{R}^n$ satisfies condition (β) if there exists a real number $\beta(T) > 0$ such that for for all $x \in K$ with $\|x\| \geq 1$, we have $\|T(x)\| \leq \beta(T) \|x\|$.

Examples

1. Any linear continuous operator $T : \mathbb{R}^n \to \mathbb{R}$ satisfies condition (β).

2. If $T : K \to \mathbb{R}^n$ is a k-Lipschitz mapping then T satisfies condition (β) with $\beta(T) = k + \beta_0$, where $\beta_0 = k\|x^0\| + \|T(x^0)\|$ and x^0 is an arbitrary element in K.

THEOREM 3.121 *Let $f : K \to \mathbb{R}^n$ be a continuous function and $T : K \to \mathbb{R}^n$ a function satisfying condition (β). If the following conditions are satisfied,*

1. $\liminf_{\|x\|\to\infty}\langle f(x) - T(x), x\rangle/\|x\|^2 \geq k_0 > 0$;

2. $\beta(T) < k_0$;

then there exists $\rho > 0$ such that $\langle x, f(x)\rangle > 0$, for all $x \in K$ with $\|x\| = \rho$.

Proof. Let $\varepsilon > 0$ be such that $\beta(T) + \varepsilon < k_0$. From assumption 1. we have that there exists $\rho_0 > 0$ such that for all $x \in K$ with $\|x\| > \rho_0$ we have

$$\frac{\langle f(x) - T(x), x\rangle}{\|x\|^2} > k_0 - \varepsilon,$$

which implies

$$\langle f(x) - T(x), x\rangle > (k_0 - \varepsilon)\|x\|^2$$

and finally,

$$\langle f(x), x\rangle > \langle T(x), x\rangle + (k_0 - \varepsilon)\|x\|^2.$$

From the last inequality we obtain

$$\langle f(x), x\rangle \geq -\beta(T)\|x\|^2 + (k_0 - \varepsilon)\|x\|^2 = \|x\|^2(-\beta(T) - \varepsilon + k_0) > 0,$$

for all $x \in K$ with $\|x\| > \rho_0$. If we take $\rho > \rho_0$, the proof is complete. $\square$

REMARK 3.10 *Theorem 3.121 is applicable in the following cases.*

(i) $f(x) = T(x) + ax + b$, *where* $a > 0$, $b \in \mathbb{R}^n$ *is an arbitrary vector, and* T *satisfies condition* (β) *with* $\beta(T) < a$.

(ii) $f(x) = T(x) + L(x) + b$, *where* $b \in \mathbb{R}^n$ *is an arbitrary vector,* L *is a linear operator from* $\mathbb{R}^n$ *into* $\mathbb{R}^n$ *such that* $\langle L(x), x\rangle \geq k_0\|x\|^2$, *for any* $x \in K$, *and* T *satisfies condition* (β) *with* $\beta(T) < k_0$.

In the n-dimensional Euclidean space Theorem 3.120 has the following form.

THEOREM 3.122 *Let* $K \subset \mathbb{R}^n$ *be a closed convex cone and* $f : \mathbb{R}^n \to \mathbb{R}^n$ *a continuous function. If there exists* $\rho > 0$ *such that* $\langle x, f(x)\rangle \geq 0$ *for any* $x \in K$ *with* $\|x\| = \rho$, *then the problem* $\mathrm{NCP}(f, K)$ *has a solution* x^* *such that* $\|x^*\| \leq \rho$.

Proof. Let T_ρ be the radial retraction onto the ball $B(0, \rho) = \{x \in \mathbb{R}^n : \|x\| \leq \rho\}$; that is,

$$T_\rho(x) = \begin{cases} x, & \text{if } \|x\| \leq \rho \\ \dfrac{\rho x}{\|x\|}, & \text{if } \|x\| > \rho. \end{cases}$$

It is known that T_ρ is continuous. If we denote $K_\rho = B(0, \rho) \cap K$, we have that T_ρ is also a continuous retraction of the cone K onto K_ρ.

We denote by $F : \mathbb{R}^n \to \mathbb{R}^n$ the continuous mapping defined by

$$F(x) = f(T_\rho(x)) + \|x - T_\rho(x)\|x.$$

For any $x \in K$ with $\|x\| > \rho$ we have $\langle x, F(x) \rangle > 0$. Indeed, we have

$$\langle x, F(x) \rangle = \langle x, f(T_\rho(x)) \rangle + \|x - T_\rho(x)\| \cdot \|x\|^2$$

$$= \left\langle \frac{\|x\|}{\rho} T_\rho(x), f(T_\rho(x)) \right\rangle + \|x - T_\rho(x)\| \cdot \|x\|^2 > 0.$$

Because the fact that $\langle x, F(x) \rangle > 0$ for any $x \in K$ with $\|x\| > \rho$, it is easy to show that F satisfies condition Θ with respect to K. Hence, by Theorem 3.118, F is without EFE with respect to K. Applying Theorem 3.116, we have that $\mathrm{NCP}(F, K)$ has a solution x^* which must satisfy the inequality $\|x^*\| \leq \rho$. Therefore $F(x^*) = f(x^*)$ and x^* is a solution to $\mathrm{NCP}(f, K)$. $\square$

REMARK 3.11 *Theorem 3.122 is known in complementarity theory, but our proof presented here is different from other proofs.*

REMARK 3.12 *If $f : \mathbb{R}^n \to \mathbb{R}^n$ is a continuous function such that there exists $\rho > 0$ with the property that $\langle x, f(x) \rangle \geq 0$ for any $x \in K$ with $\|x\| = \rho$, then the function $F(x) = f(T_\rho(x)) + \|x - T_\rho(x)\|x$ is such that*

$$\liminf_{\substack{\|x\| \to +\infty \\ x \in K}} \langle x, F(x) \rangle = +\infty.$$

Indeed, we have (for any $x \in K$ with $\|x\| > \rho$)

$$\langle x, F(x) \rangle = \langle x, f(T_\rho(x)) \rangle + \|x - T_\rho(x)\| \cdot \|x\|^2$$

$$= \frac{\|x\|}{\rho} \langle T_\rho(x), f(T_\rho(x)) \rangle + \|x - T_\rho(x)\| \cdot \|x\|^2$$

$$\geq \|x - T_\rho(x)\| \cdot \|x\|^2 \geq \rho^2 \|x - T_\rho(x)\| \geq \rho^2 [\|x\| - \rho].$$

Computing $\liminf_{\substack{\|x\| \to +\infty \\ x \in K}}$, we obtain that

$$\liminf_{\substack{\|x\| \to +\infty \\ x \in K}} \langle x, F(x) \rangle = +\infty.$$

$\square$

3.9.3   The Asymptotic Browder–Hartman–Stampacchia Condition

Let $(\mathbb{R}^n, \langle \cdot, \cdot \rangle)$ be the n-dimensional Euclidean space, $K \subset \mathbb{R}^n$ a closed pointed convex cone, and $f : \mathbb{R}^n \to \mathbb{R}^n$ a continuous function. We introduce the following condition.

DEFINITION 3.123 *We say that f satisfies the* asymptotic Browder–Hartman–Stampacchia condition *(denoted ABHS) with respect to K if*

$$\liminf_{\substack{\|x\|\to+\infty \\ x\in K}} \langle x, f(x)\rangle = +\infty.$$

The relations between conditions BHS and ABHS are given by the following result.

PROPOSITION 3.8 *Let $K \subset \mathbb{R}^n$ be a closed pointed convex cone and $f : \mathbb{R}^n \to \mathbb{R}^n$ a continuous function. If f satisfies condition ABHS, then f satisfies condition BHS. If f satisfies condition BHS, then the function $F(x) = f(T_\rho(x)) + \|x - T_\rho(x)\|x$ satisfies condition ABHS.*

Proof. We suppose that f satisfies condition ABHS; that is, we have that

$$\liminf_{\substack{\|x\|\to+\infty \\ x\in K}} \langle x, f(x)\rangle = +\infty.$$

In this case given $r > 0$, there exists $\rho > 0$ such that for any $x \in K$ with $\|x\| = \rho$ we have $\langle x, f(x)\rangle > r$. Indeed, if this is not true, then for any $n \in \mathbb{N}$, there exists $x_n \in K$ with $\|x_n\| = n$ such that $\langle x_n, f(x_n)\rangle \leq r$. Therefore, condition ABHS is not satisfied. This contradiction implies that f satisfies condition BHS. Conversely, if f satisfies condition BHS, then by Remark 3.12 F satisfies condition ABHS. $\qquad\square$

COROLLARY 3.124 *If $K \subset \mathbb{R}^n$ is a closed pointed convex cone and $f : \mathbb{R}^n \to \mathbb{R}^n$ is a continuous function satisfying condition ABHS, then $\mathrm{NCP}(f, K)$ has a solution.*

Proof. This corollary is a consequence of Proposition 3.8 and Theorem 3.122. $\qquad\square$

In the next pages we show that condition ABHS is a good mathematical tool for the study of properties (a) and (b) of the interior band mapping $\mathcal{U}$ with respect to the cone $\mathbb{R}^n_+$.

About the existence of a solution to the $\mathrm{NCP}(f, K)$ in $\mathbb{R}^n$ we also cite the following result.

PROPOSITION 3.9 *Let $K \subset \mathbb{R}^n$ be a closed pointed convex cone and $f : \mathbb{R}^n \to \mathbb{R}^n$ a continuous function. If f satisfies the condition*

$$\liminf_{\substack{\|x\|\to\infty \\ x\in K}} \langle x, f(x)\rangle > 0, \tag{3.79}$$

then the $\mathrm{NCP}(f, K)$ has a solution.

Proof. It is sufficient to prove that f is without EFE with respect to K. Indeed, if we suppose that f has an EFE, namely, $\{x^r\}_{r>0} \subset K$, then we have

$$\langle x^r, f(x^r) \rangle = \langle x^r, u^r - \mu_r x^r \rangle$$
$$= \langle x^r, u^r \rangle - \mu_r \langle x^r, x^r \rangle = -\mu_r \|x^r\|^2 < 0,$$

which implies

$$\liminf_{\|x^r\| \to \infty} \langle x^r, f(x^r) \rangle \leq 0.$$

This relation is impossible because we supposed condition (3.79). $\square$

THEOREM 3.125 *If $f : \mathbb{R}^n_+ \to \mathbb{R}^n$ is a continuous function, then*

$$\liminf_{\substack{\|x\| \to \infty \\ x \in \mathbb{R}^n_{++}}} \langle x, f(x) \rangle = +\infty$$

if and only if

$$\liminf_{\substack{\|x\| \to \infty \\ x \in \mathbb{R}^n_+}} \langle x, f(x) \rangle = +\infty.$$

Proof. Obviously if

$$\liminf_{\substack{\|x\| \to \infty \\ x \in \mathbb{R}^n_+}} \langle x, f(x) \rangle = +\infty,$$

then

$$\liminf_{\substack{\|x\| \to \infty \\ x \in \mathbb{R}^n_{++}}} \langle x, f(x) \rangle = +\infty.$$

The converse follows if we show that if

$$\liminf_{\substack{\|x\| \to \infty \\ x \in \mathbb{R}^n_{++}}} \langle x, f(x) \rangle = +\infty,$$

then

$$\liminf_{\substack{\|x\| \to \infty \\ x \in \partial \mathbb{R}^n_+}} \langle x, f(x) \rangle = +\infty.$$

Let $\{x^n\}$ be a sequence such that $\|x^n\| \to \infty$ as $n \to \infty$ and for any $n \in \mathbb{N}$, $x^n \in \partial \mathbb{R}^n_+$. Let n be fixed. Because $x^n \in \partial \mathbb{R}^n_+$, there exists $\{y^m\} \subset \mathbb{R}^n_{++}$ such that $\{y^m\} \to x^n$ as $m \to \infty$. For any $n \in \mathbb{N}$ we select such a sequence $\{y^m\}$. Let $\varepsilon_0 > 0$ be an arbitrary real number. For any $n \in \mathbb{N}$ we can select the sequence $\{y^m\}$ such that $\|y^m - x^n\| < \varepsilon_0$, for any $m \in \mathbb{N}$. We can suppose that for any $n \in \mathbb{N}$, $\|f(x^n)\| > 0$. Because $\lim_{m \to \infty} \|x^n - y^m\| = 0$, there exists $m_1 \in \mathbb{N}$ such that

$$\|x^n - y^m\| < \frac{1}{\|f(x^n)\|},$$

for any $m > m_1$.

Because f is continuous, there exists $m_2 \in \mathbb{N}$ such that

$$\|f(x^n) - f(y^m)\| < \frac{1}{\varepsilon_0 + \|x^n\|},$$

for any $m > m_2$. For any $m > \max\{m_1, m_2\}$ we have

$$\|y^m\| \leq \|x^n - y^m\| + \|x^n\| < \varepsilon_0 + \|x^n\|$$

and

$$\|f(x^n) - f(y^m)\| < \frac{1}{\varepsilon_0 + \|x^n\|} < \frac{1}{\|y^m\|}.$$

We take an arbitrary $m > \max\{m_1, m_2\}$ and we have

$$\|x^n - y^m\| < \frac{1}{\|f(x^n)\|}$$

and

$$\|f(x^n) - f(y^m)\| < \frac{1}{\|y^m\|}.$$

If we denote $y^{n,*} = y^m$ then we have

$$\|x^n - y^{n,*}\| < \frac{1}{\|f(x^n)\|}$$

and

$$\|f(x^n) - f(y^{n,*})\| < \frac{1}{\|y^{n,*}\|}.$$

We have

$$\langle y^{n,*}, f(y^{n,*}) \rangle = [\langle y^{n,*}, f(y^{n,*}) \rangle - \langle x^n, f(x^n) \rangle] + \langle x^n, f(x^n) \rangle$$

and

$$\|\langle y^{n,*}, f(y^{n,*}) \rangle - \langle x^n, f(x^n) \rangle\|$$
$$\leq \|\langle x^n - y^{n,*}, f(x^n) \rangle\| + \|\langle y^{n,*}, f(x^n) - f(y^{n,*}) \rangle\|$$
$$< \frac{1}{\|f(x^n)\|} \cdot \|f(x^n)\| + \|y^{n,*}\| \cdot \frac{1}{\|y^{n,*}\|} = 2.$$

Therefore,

$$\langle y^{n,*}, f(y^{n,*}) \rangle < 2 + \langle x^n, f(x^n) \rangle. \tag{3.80}$$

Because $\|x^n\| - \|y^{n,*}\| \leq \|x^n - y^{n,*}\| < \varepsilon_0$, for any $n \in \mathbb{N}$, we have $\|x^n\| < \varepsilon_0 + \|y^{n,*}\|$, for any $n \in \mathbb{N}$, which implies $\|y^{n,*}\| \to +\infty$, because $\|x^n\| \to +\infty$.

Computing $\lim\inf$ in (3.80) we have

$$\liminf_{n \to \infty} \langle x^n, f(x^n) \rangle = +\infty.$$

Therefore,

$$\liminf_{\substack{\|x\| \to +\infty \\ x \in \partial \mathbb{R}_+^n}} \langle x, f(x) \rangle = +\infty$$

and the proof is complete.   □

3.9.4   Infinitesimal Interior-Point-ε-Exceptional Families

DEFINITION 3.126 *Let* $g : \mathbb{R}^n \to \mathbb{R}^n$ *be a function. Given a scalar* $\varepsilon > 0$, *we say that a sequence* $\{y^r\}_{r>0} \subset \mathbb{R}_{++}^n$ *is an* infinitesimal interior-point-ε-exceptional family *for* g *if* $\|y^r\| \to 0$ *as* $r \to +\infty$ *and for each* y^r *there exists a positive number* $0 < \mu_r < 1$ *such that*

$$g_l(y^r) = \frac{1}{2}\left(\mu_r - \frac{1}{\mu_r}\right) y_l^r + \frac{\varepsilon\mu_r}{y_l^r}\|y^r\|^4, \tag{3.81}$$

for all $l = 1, 2, \ldots, n$.

PROPOSITION 3.10 *If* $f : \mathbb{R}^n \to \mathbb{R}^n$ *is a continuous function and* $g = \mathcal{I}(f)$ *is the inversion of* f, *then* $\{x^r\} \subset \mathbb{R}_{++}^n$ *is an interior-point-ε-exceptional family for* f *if and only if* $\{y^r\} \subset \mathbb{R}_{++}^n$ *is an infinitesimal interior-point-ε-exceptional family for* g, *where* $y^r = i(x^r)$ *is the inversion of* x^r, *for all* $r > 0$.

Proof. Suppose that $\{x^r\}_{r>0} \subset \mathbb{R}_{++}^n$ is an interior-point-ε-exceptional family for f and let

$$y^r = i(x^r), \tag{3.82}$$

for all $r > 0$. Because $i^{-1} = i$, Equations (3.76) and (3.82) imply that

$$f_l(i(y^r)) = \frac{1}{2}\left(\mu_r - \frac{1}{\mu_r}\right)i(y^r)_l + \frac{\varepsilon\mu_r}{i(y^r)_l}, \tag{3.83}$$

for all $l = 1, 2, \ldots, n$. Multiplying both sides of Equation (3.83) by $\|y^r\|^2$ we obtain Equation (3.81). Hence, $\{y^r\}_{r>0} \subset \mathbb{R}_{++}^n$ is an infinitesimal interior-point-ε-exceptional family for g. Similarly, it can be proved that if $\{y^r\}_{r>0} \subset \mathbb{R}_{++}^n$ is an infinitesimal interior-point-ε-exceptional family for g, then $\{x^r\}_{r>0} \subset \mathbb{R}_{++}^n$ is an interior-point-ε-exceptional family for f.   □

THEOREM 3.127 *Let $f : \mathbb{R}^n \to \mathbb{R}^n$ be a continuous function and $\varepsilon > 0$. If there is no infinitesimal interior-point-ε-exceptional family for $g = \mathcal{I}(f)$, then there exists a point $x(\varepsilon)$ such that*

$$x(\varepsilon) > 0, \; f(x(\varepsilon)) > 0, \; x_l(\varepsilon) f_l(x(\varepsilon)) = \varepsilon, \tag{3.84}$$

for all $l = 1, 2, \ldots, n$.

Proof. Suppose to the contrary, that there is no point $x(\varepsilon)$ which satisfies relation (3.84). Then, by Theorem 3.114, the function f has an interior-point-ε-exceptional family $\{x^r\}_{r>0} \subset \mathbb{R}^n_{++}$. Hence, Proposition 3.10 implies that $\{y^r\}_{r>0} \subset \mathbb{R}^n_{++}$ is an infinitesimal interior-point-ε-exceptional family for g, where $y^r = i(x^r)$, for all $r > 0$. But this is in contradiction with our assumption. $\qquad \square$

By Theorem 3.127 it is interesting to find conditions under which the inversion of a continuous function does not possess an infinitesimal interior-point-ε-exceptional family for all $\varepsilon > 0$. For such functions $\mathcal{U}(\varepsilon) \neq \emptyset$, for each $\varepsilon > 0$.

3.9.5 Results Related to Properties (a) and (b) of the Interior Band Mapping $\mathcal{U}$

THEOREM 3.128 *Let $f : \mathbb{R}^n \to \mathbb{R}^n$ be a continuous function. If*

$$\liminf_{\substack{\|x\| \to +\infty \\ x \in \mathbb{R}^n_{++}}} \langle x, f(x) \rangle = +\infty,$$

then

1. *The problem $\mathrm{NCP}(f, \mathbb{R}^n_+)$ has a solution.*

2. *$\mathcal{U}(\varepsilon) \neq \emptyset$, for any $\varepsilon > 0$.*

3. *For any fixed $\varepsilon_0 > 0$ the set $\cup_{\varepsilon \in]0,\varepsilon_0]} \mathcal{U}(\varepsilon)$ is bounded.*

Proof.

1. By Theorem 3.125 we have that

$$\liminf_{\substack{\|x\| \to +\infty \\ x \in \mathbb{R}^n_+}} \langle x, f(x) \rangle = +\infty$$

   and by Proposition 3.8 we have that f satisfies condition BHS. Applying Theorem 3.122 we obtain that $\mathrm{NCP}(f, \mathbb{R}^n_+)$ has a solution.

2. By using Theorem 3.114 it is sufficient to show that f does not have an interior-point-ε-exceptional family $\{x^r\}_{r>0} \subset \mathbb{R}^n_{++}$. Indeed, we suppose

that f has an interior-point-ε-exceptional family $\{x^r\}_{r>0} \subset \mathbb{R}^n_{++}$. Multiplying formula (3.76) given in Definition 3.113 by x_l^r and summing with l from 1 to n we obtain

$$\langle x^r, f(x^r) \rangle = \frac{1}{2} \left(\mu_r - \frac{1}{\mu_r} \right) \|x^r\|^2 + n\varepsilon\mu_r,$$

where $0 < \mu_r < 1$, for any $r > 0$. From the last equality we deduce

$$\langle x^r, f(x^r) \rangle + \frac{1}{2} \left(\frac{1}{\mu_r} - \mu_r \right) \|x^r\|^2 < n\varepsilon.$$

Let $r_0 > 0$ such that $\|x^{r_0}\| > 0$. Because $\|x^r\| \to +\infty$ as $r \to +\infty$, we can consider a subsequence $\{x^{r_i}\}$ such that $\|x^{r_0}\| < \|x^{r_i}\|$ and $\|x^{r_i}\| \to +\infty$ as $i \to \infty$. For this subsequence we have

$$\frac{1}{2} \left(\frac{1}{\mu_{r_i}} - \mu_{r_i} \right) \|x^{r_0}\|^2 + \langle x^{r_i}, f(x^{r_i}) \rangle < n\varepsilon.$$

Computing lim inf and using the assumption of our theorem, we obtain a contradiction. Therefore by Theorem 3.114 we have that $\mathcal{U}(\varepsilon) \neq \emptyset$, for any $\varepsilon > 0$.

3. We observe that for any $x(\varepsilon) \in \mathcal{U}(\varepsilon)$ we have $\langle x(\varepsilon), f(x(\varepsilon)) \rangle = n\varepsilon$. Now, we suppose that there is an $\varepsilon > 0$ such that $\cup_{\varepsilon \in]0,\varepsilon_0]} \mathcal{U}(\varepsilon)$ is not bounded. Then, there is a sequence $x(\varepsilon_k) \in \mathcal{U}(\varepsilon_k)$ with $\varepsilon_k \in]0, \varepsilon_0]$ which is not bounded. Hence, by the assumption of our theorem we have

$$\liminf_{k \to \infty} \langle x(\varepsilon_k), f(x(\varepsilon_k)) \rangle = +\infty.$$

On the other hand,

$$\langle x(\varepsilon_k), f(x(\varepsilon_k)) \rangle = n\varepsilon_k \leq n\varepsilon_0,$$

which implies

$$\liminf_{k \to \infty} \langle x(\varepsilon_k), f(x(\varepsilon_k)) \rangle \leq n\varepsilon_0$$

and we have a contradiction. Therefore $\cup_{\varepsilon \in]0,\varepsilon_0]} \mathcal{U}(\varepsilon)$ is bounded for all $\varepsilon_0 > 0$.

$\square$

We remark that in Theorem 3.128 conclusion 1 is (a) of $\mathcal{U}$ and conclusion 2 is property (b).

THEOREM 3.129 *Let* $f : \mathbb{R}^n \to \mathbb{R}^n$ *be a continuous function. If*

$$\liminf_{\substack{\|x\| \to +\infty \\ x \in \mathbb{R}^n_{++}}} \frac{\langle x, f(x) \rangle}{\|x\|^2} > 0,$$

then the interior band mapping $\mathcal{U}$ *has properties (a) and (b).*

Proof. We denote

$$r = \liminf_{\substack{\|x\| \to +\infty \\ x \in \mathbb{R}^n_{++}}} \frac{\langle x, f(x) \rangle}{\|x\|^2}$$

and we take r_0 such that $0 < r_0 < r$. There exists $\rho > 0$ such that for any $x \in \mathbb{R}^n_{++}$ with $\|x\| > \rho$,

$$\frac{\langle x, f(x) \rangle}{\|x\|^2} > r_0.$$

Indeed if this is not true, then for any $n \in \mathbb{N}$, there exists $x^n \in \mathbb{R}^n_{++}$ with $\|x^n\| > n$ and such that

$$\frac{\langle x^n, f(x^n) \rangle}{\|x^n\|^2} \leq r_0.$$

Because $\|x^n\| \to +\infty$ as $n \to \infty$ we have that

$$\liminf_{n \to \infty} \frac{\langle x^n, f(x^n) \rangle}{\|x^n\|^2} \leq r_0 < r,$$

which is impossible. Therefore, for any $x \in \mathbb{R}^n_{++}$ with $\|x\| > \rho$ we have $\langle x, f(x) \rangle > r_0 \|x\|^2$ which implies that

$$\liminf_{\substack{\|x\| \to +\infty \\ x \in \mathbb{R}^n_{++}}} \langle x, f(x) \rangle = +\infty.$$

Applying Theorem 3.128 we obtain that the interior band mapping $\mathcal{U}$ has properties (a) and (b). $\qquad\square$

THEOREM 3.130 *Let* $f : \mathbb{R}^n \to \mathbb{R}^n$ *be a continuous function and* $g = \mathcal{I}(f)$. *If the lower scalar derivative of* g *in* 0 *along* $\mathbb{R}^n_{++}$ *is positive, then the interior band mapping* $\mathcal{U}$ *has properties (a) and (b).*

Proof. We have

$$g^{\#}(0) = \liminf_{\substack{y \to 0 \\ y \in \mathbb{R}^n_{++}}} \frac{\langle g(y), y \rangle}{\|y\|^2}. \tag{3.85}$$

Let $y = i(x)$. Then, we have

$$\liminf_{\substack{y \to 0 \\ y \in \mathbb{R}^n_{++}}} \frac{\langle g(y), y \rangle}{\|y\|^2} = \liminf_{\substack{\|x\| \to \infty \\ x \in \mathbb{R}^n_{++}}} \frac{\langle f(x), x \rangle}{\|x\|^2}. \tag{3.86}$$

Equations (3.85) and (3.86) imply

$$\underline{g}^{\#}(0) = \liminf_{\substack{\|x\| \to \infty \\ x \in \mathbb{R}^n_{++}}} \frac{\langle f(x), x \rangle}{\|x\|^2}.$$

Hence, the result follows by using Theorem 3.128.   □

The following definition extends the notion of co-positive functions (see [Isac, 2000d]). We remark that the notion can be introduced along an arbitrary convex cone in a Hilbert space too, but for our investigation it is sufficient to consider the case of $\mathbb{R}^n$ with the cone $\mathbb{R}^n_{++}$.

DEFINITION 3.131 *The function* $f : \mathbb{R}^n \to \mathbb{R}^n$ *is called* asymptotically co-positive *along* $\mathbb{R}^n_{++}$ *if there is a* $\rho > 0$ *such that*

$$\langle f(x), x \rangle \geq 0,$$

for all $x \in \mathbb{R}^n_{++}$ *with* $\|x\| > \rho$.

The following definition can be formulated along an arbitrary convex cone in a Hilbert space too, but for the same reason as above we would consider the case of $\mathbb{R}^n$ with the cone $\mathbb{R}^n_{++}$ only.

DEFINITION 3.132 *The function* $f : \mathbb{R}^n \to \mathbb{R}^n$ *is called* strongly asymptotically co-positive *along* $\mathbb{R}^n_{++}$ *if there are* $\beta, \rho > 0$ *such that*

$$\langle f(x), x \rangle \geq \beta \|x\|^2,$$

for all $x \in \mathbb{R}^n_{++}$ *with* $\|x\| > \rho$.

We remark that f is strongly asymptotically co-positive along $\mathbb{R}^n_{++}$ if and only if there is a $\beta > 0$ such that the function $f - \beta I$ is asymptotically co-positive along $\mathbb{R}^n_{++}$, where I is the identity function of $\mathbb{R}^n$.

The following theorem follows directly from Theorem 3.128 and Definition 3.132.

THEOREM 3.133 *If* $f : \mathbb{R}^n \to \mathbb{R}^n$ *is a continuous strongly asymptotically co-positive function along* $\mathbb{R}^n_{++}$, *then the interior band mapping* $\mathcal{U}$ *has properties* (a) *and* (b).

COROLLARY 3.134 *Let* $f : \mathbb{R}^n \to \mathbb{R}^n$ *be a continuous function. If there is a* $\rho > 0$ *and* $\beta > 0$ *such that* $f(x) - \beta x \in \mathbb{R}^n_+$, *for all* $x \in \mathbb{R}^n_{++}$ *with* $\|x\| > \rho$, *then the interior band mapping* $\mathcal{U}$ *has properties* (a) *and* (b).

Proof. We have

$$\langle f(x) - \beta x, x \rangle \geq 0$$

for all $x \in \mathbb{R}^n_{++}$ with $\|x\| > \rho$. Hence,

$$\langle f(x), x \rangle \geq \beta \|x\|^2,$$

for all $x \in \mathbb{R}^n_{++}$ with $\|x\| > \rho$. Therefore, f is strongly asymptotically co-positive along $\mathbb{R}^n_{++}$ and the result follows from Theorem 3.133. □

At the end of our section we present two results: Theorem 3.136 and Corollary 3.137, which show that the coercivity condition of Theorem 3.129 can be satisfied by a large class of functions. For this we need the following corollary which is a particularization of Corollary 3.1 [Isac and Nemeth, 2003]:

COROLLARY 3.135 *Let* $D = \{x \in \mathbb{R}^n : \|x\| \leq 1\}$ *and* $f : \mathbb{R}^n \to \mathbb{R}^n$; $f(0) = 0$. *There exists a unique extension* $\tilde{f} : \mathbb{R}^n \to \mathbb{R}^n$ *of* $f|_D$ *such that* $\tilde{f}$ *is a fixed point of* $\mathcal{I}$ *(i.e.,* $\tilde{f} = \mathcal{I}(\tilde{f})$).

We proved in our paper [Isac and Nemeth, 2003] that this extension has the form

$$\tilde{f}(x) = \begin{cases} f(x) & \text{if } \|x\| \leq 1 \\ \mathcal{I}(f)(x) & \text{if } \|x\| > 1 \end{cases},$$

Because $\tilde{f} = \mathcal{I}(\tilde{f})$ and

$$\liminf_{\|x\| \to \infty} \frac{\langle \tilde{f}(x), x \rangle}{\|x\|^2} = \underline{\mathcal{I}(\tilde{f})}^{\#}(0) \text{ (by Lemma 4.1 of [Isac and Nemeth, 2003]),}$$

we have

$$\liminf_{\|x\| \to \infty} \frac{\langle \tilde{f}(x), x \rangle}{\|x\|^2} = \underline{\tilde{f}}^{\#}(0) = \underline{f}^{\#}(0).$$

Hence,

$$\liminf_{\|x\| \to \infty} \frac{\langle \tilde{f}(x), x \rangle}{\|x\|^2} > 0 \tag{3.87}$$

if and only if

$$\underline{f}^{\#}(0) > 0.$$

On the other hand, by using the Cauchy inequality it is easy to see that if $\tilde{f}$ satisfies condition (3.87), then for any $b \in \mathbb{R}^n$, $\tilde{f} + b$ also satisfies condition (3.87). Hence, we have the following result.

THEOREM 3.136 *Let* $b \in \mathbb{R}^n$, $f : \mathbb{R}^n \to \mathbb{R}^n$ *with* $f(0) = 0$ *and* $\underline{f}^{\#}(0) > 0$, $\tilde{f} : \mathbb{R}^n \to \mathbb{R}^n$ *with*

$$\tilde{f}(x) = \begin{cases} f(x) & \text{if } \|x\| \leq 1 \\ \mathcal{I}(f)(x) & \text{if } \|x\| > 1 \end{cases}$$

and $F = \tilde{f} + b$. Then, F satisfies the condition

$$\liminf_{\|x\| \to \infty} \frac{\langle F(x), x \rangle}{\|x\|^2} > 0.$$

COROLLARY 3.137 *Let* $a, b \in \mathbb{R}^n$, $f : \mathbb{R}^n \to \mathbb{R}^n$ *with* $f(0) = a$ *and* $\underline{f}^{\#}(0) > 0$, *and* $F : \mathbb{R}^n \to \mathbb{R}^n$ *with*

$$F(x) = \begin{cases} f(x) - a + b & \text{if } \|x\| \leq 1 \\ \mathcal{I}(f)(x) - \|x\|^2 a + b & \text{if } \|x\| > 1 \end{cases}.$$

Then, F satisfies the condition

$$\liminf_{\|x\| \to \infty} \frac{\langle F(x), x \rangle}{\|x\|^2} > 0.$$

Proof. Let $f_0(x) = f(x) - a$. It is easy to see that $f_0(0) = 0$ and $\underline{f_0}^{\#}(0) = \underline{f}^{\#}(0) > 0$. Hence, we can apply Theorem 3.136 to the function f_0 to obtain the desired result.   □

For $A : \mathbb{R}^n \to \mathbb{R}^n$ a linear operator, we denote by A_s the operator $(A + A^*)/2$, where A^* is the adjoint of A. Let $\sigma(A_s)$ be the spectrum of A_s. With these notations we have as follows.

REMARK 3.13 *If f satisfies supplementary conditions in 0, [Nemeth, 1993] provides useful computational formulae for checking the condition $\underline{f}^{\#}(0) > 0$:*

1. *By Theorem 1.1 of [Nemeth, 1993] if f is locally Lipschitz in 0 and the directional derivative $f'(0; h)$ exists for each h, then*

$$\underline{f}^{\#}(0) = \inf_{\|h\|=1} \langle f'(0; h), h \rangle.$$

2. *By Theorem 1.2 of [Nemeth, 1993] if f is Frechét differentiable in 0, with the differential $df(0)$, then*

$$\underline{f}^{\#}(0) = \min_{\|h\|=1} \langle df(0)(h), h \rangle.$$

3. *By Theorem 1.5 of [Nemeth, 1993] if f is Frechét differentiable in 0 with the differential $df(0)$, then*

$$\underline{f}^{\#}(0) = \min \sigma((df(0)))_s.$$

3.9.6   Comments

In this section we studied the interior band of ε-solutions of the nonlinear complementarity problem defined by a continuous function from $\mathbb{R}^n$ to $\mathbb{R}^n$ and by the cone $\mathbb{R}^n_+$.

By the results presented in this section we put in evidence the importance of the asymptotic Browder–Hartman–Stampacchia condition. By using this condition and the scalar derivative we obtained some new results related to the interior band of ε-solutions.

By our method we do not need to suppose that the mapping f is uniformly P-mapping or monotone mapping as in several papers cited in our references.

Our ideas presented in this section may be a starting point for new developments.

3.10   REFE-acceptable Mappings and a Necessary and Sufficient Condition for the Nonexistence of Regular Exceptional Families of Elements

A topological method, now developing in complementarity theory, is based on the notion of an exceptional family of elements introduced by the topological degree in 1997 in [Isac et al., 1997], for completely continuous fields (see also [Bulavski et al., 1998]). By using Leray–Schauder type alternatives, the extension of this notion to other classes of mappings can be successfully applied to proving existence theorems for complementarity problems [Isac, 2000b; Isac, 2006; Isac and Carbone, 1999; Isac and Kalashnikov, 2001]. It is known that if a completely continuous field (or a k-set-contraction field) is without EFE, then the complementarity problem, associated with this field and with a closed convex cone in a Hilbert space, has a solution. The section has two goals. The first goal is to show that the notion of EFE for more general classes of mappings than those based on Leray–Schauder type alternatives can also be successfully applied to complementarity problems in a similar manner as described above. This motivates the introduction of the notion of REFE-acceptable mappings. The second goal is to present necessary and sufficient conditions for the nonexistence of a regular exceptional family of elements. We note that our condition is the first known necessary and sufficient condition in the literature related to the nonexistence of EFE. By this condition we obtain new and interesting existence theorems for nonlinear and linear complementarity problems. If the complementarity problem is described by a pseudo-monotone mapping some of these theorems become existence and uniqueness theorems.

3.10.1   REFE-Acceptable Mappings

Let $(H, \langle \cdot, \cdot \rangle)$ be a Hilbert space, $K \subset H$ a closed convex cone, and $f : H \to H$ a mapping. For any $r > 0$ we denote $K_r = \{x \in K \mid \|x\| \leq r\}$ and P_{K_r} is the projection mapping onto K_r. We recall the definition of the notion of regular exceptional family of elements [Isac et al., 1997].

DEFINITION 3.138 *Let* $(H, \langle \cdot, \cdot \rangle)$ *be a Hilbert space,* $K \subset H$ *a closed convex cone, and* $f : H \to H$ *a mapping. We say that a family of elements* $\{x_r\}_{r>0} \subset K$ *is a* regular exceptional family of elements *(denoted* REFE*) for* f *with respect to* K, *if for every real number* $r > 0$, *there exists a real number* $\mu_r > 0$ *such that the vector* $u_r = \mu_r x_r + f(x_r)$ *satisfies the following conditions:*

1. $u_r \in K^*$.

2. $\langle u_r, x_r \rangle = 0$.

3. $\|x_r\| = r$.

DEFINITION 3.139 *Let* $(H, \langle \cdot, \cdot \rangle)$ *be a Hilbert space and* $K \subset H$ *a closed convex cone. A mapping* $f : H \to H$ *is called* REFE-acceptable *if either the problem* $\mathrm{NCP}(f, K)$ *has a solution, or the mapping* f *has a REFE with respect to* K.

Obviously we have the following.

LEMMA 3.140 *Let* $(H, \langle \cdot, \cdot \rangle)$ *be a Hilbert space,* $K \subset H$ *a closed convex cone, and* $f : H \to H$ *a REFE-acceptable mapping with respect to* K. *If* f *is without a REFE with respect to* K, *then the problem* $\mathrm{NCP}(f, K)$ *has a solution.*

PROPOSITION 3.11 *[Bianchi et al., 2004] If there exists* $x_* \in K_r$ *such that* $\langle f(x_*), x - x_* \rangle \geq 0$ *for any* $x \in K_r$ *and there exists* $y \in K_r$ *with* $\|y\| < r$ *such that* $\langle f(x_*), x_* - y \rangle \geq 0$, *then we have* $\langle f(x_*), x - x_* \rangle \geq 0$ *for any* $x \in K$.

Proof. We consider the convex continuous mapping $\phi(x) = \langle f(x_*), x - x_* \rangle$, defined for any $x \in K$. We have $\phi(x) \geq 0$ for any $x \in K_r$ and $\phi(x_*) = 0$. Then, x_* is a global minimum of ϕ on K_r. Because we have $0 \leq \phi(y) = \langle f(x_*), y - x_* \rangle \leq 0 = \phi(x_*)$, we deduce that y is also a global minimum of ϕ on K_r. Therefore, (because $\|y\| < r$) we have that y is a local minimum of ϕ on K and because ϕ is convex y is a global minimum on K. Because $\phi(y) = \phi(x_*)$ we obtain that x_* is a global minimum of ϕ on K; that is, we have $\langle f(x_*), x - x_* \rangle \geq 0$ for any $x \in K$. ☐

We have the following result.

THEOREM 3.141 *Let* $(H, \langle \cdot, \cdot \rangle)$ *be a Hilbert space,* $K \subset H$ *a closed convex cone, and* $f : H \to H$ *a mapping such that for any* $r > 0$ *the mapping* $\Psi_r = P_{K_r} \circ (I - f)$ *has a fixed point (which is necessarily an element of* K_r), *where the mapping Ψ_r is considered from K_r into K_r. Then f is a REFE-acceptable mapping with respect to K.*

Proof. If the problem $\mathrm{NCP}(f, K)$ has a solution we have nothing to prove. We suppose that the problem $\mathrm{NCP}(f, K)$ has no solution. In this case we show that f has a REFE with respect to K. For every $r > 0$ there exists $x_r \in K_r$ such that

$$x_r = \Psi_r(x_r) = P_{K_r}(x_r - f(x_r)).$$

We know (see [Isac, 2006], Chapter 2) that in this case we have

$$\langle f(x_r), x - x_r \rangle \geq 0 \text{ for any } x \in K_r. \tag{3.88}$$

(Because we supposed that $\mathrm{NCP}(f, K)$ has no solution, we have that (3.88) is not satisfied for any $x \in K$.)

We show following the ideas of Bianchi, Hadjisavvas, and Schaible [Bianchi et al., 2004], Theorem 5.1, that $\{x_r\}_{r>0}$ is a REFE for f with respect to K.

For every $r > 0$ we define

$$\mu_r = -\frac{\langle f(x_r), x_r \rangle}{r^2}$$

and

$$u_r = \mu_r x_r + f(x_r).$$

If $\|x_r\| < r$, then taking $y = x_r$ in Proposition 3.11, we obtain that $\langle f(x_r), x - x_r \rangle \geq 0$ for any $x \in K$; that is, x_r is a solution of the $\mathrm{NCP}(f, K)$ which is impossible. Therefore, we must have $\|x_r\| = r$, for any $r > 0$. Also, we have

$$\langle x_r, u_r \rangle = \langle x_r, \mu_r x_r + f(x_r) \rangle = \langle x_r, \mu_r x_r \rangle + \langle x_r, f(x_r) \rangle$$

$$= -\left\langle x_r, \frac{\langle f(x_r), x_r \rangle}{r^2} x_r \right\rangle + \langle f(x_r), x_r \rangle = 0.$$

The number μ_r is strictly positive. Indeed, we have $\langle f(x_r), 0 - x_r \rangle \geq 0$ which implies $\langle f(x_r), x_r \rangle \leq 0$ and hence

$$\mu_r = \frac{\langle f(x_r), x_r \rangle}{r^2} \geq 0.$$

If $\mu_r = 0$, then $\langle f(x_r), x_r \rangle = 0 = \langle f(x_r), x_r - 0 \rangle$ and taking $y = 0$ in Proposition 3.11 we deduce that $\langle f(x_r), x - x_r \rangle \geq 0$ for any $x \in K$; that is, the $\mathrm{NCP}(f, K)$ has a solution which is impossible. Therefore, we have $\mu_r > 0$, for any $r > 0$. The theorem will be proved if we show that $u_r \in K^*$ for any $r > 0$. To show this, it is sufficient to prove that

$$\left\langle f(x_r), x - \frac{\langle x_r, x \rangle}{r^2} x_r \right\rangle \geq 0, \qquad (3.89)$$

for any $x \in K$.

Indeed, if (3.89) is true, then we have (because $f(x_r) = u_r - \mu_r x_r$)

$$0 \leq \left\langle u_r - \mu_r x_r, x - \frac{\langle x_r, x \rangle}{r^2} x_r \right\rangle = \langle u_r, x \rangle - \left\langle u_r, \frac{\langle x_r, x \rangle}{r^2} x_r \right\rangle - \mu_r \langle x_r, x \rangle$$

$$+ \mu_r \frac{\langle x_r, x \rangle}{r^2} r^2 = \langle u_r, x \rangle.$$

Now we show that (3.89) is true. Let $r > 0$ be fixed. We denote

$$y = x - \frac{\langle x_r, x \rangle}{r^2} x_r$$

and $z_\lambda = y + \lambda x_r$, with

$$\lambda > \frac{\langle x_r, x \rangle}{r^2}.$$

Then $z_\lambda \in K$ and

$$\frac{z_\lambda}{\|z_\lambda\|} r \in K_r.$$

Hence, we have

$$\left\langle f(x_r), \frac{z_\lambda}{\|z_\lambda\|} r - x_r \right\rangle \geq 0,$$

which implies (because $y = z_\lambda - \lambda x_r$)

$$\left\langle f(x_r), y + \left(\lambda - \frac{\|z_\lambda\|}{r} \right) x_r \right\rangle \geq 0. \qquad (3.90)$$

We can show that $\langle y, x_r \rangle = 0$, which implies that $\|z_\lambda\| = \sqrt{\|y\|^2 + \lambda^2 r^2}$. We also have

$$\lim_{\lambda \to +\infty} \left(\lambda - \frac{\|z_\lambda\|}{r} \right) = \lim_{\lambda \to +\infty} \frac{r\lambda - \|z_\lambda\|}{r}$$

$$= \lim_{\lambda \to +\infty} \frac{r^2 \lambda^2 - (\|y\|^2 + \lambda^2 r^2)}{r(r\lambda + \|z_\lambda\|)} = \lim_{\lambda \to +\infty} \frac{-\|y\|^2}{r^2(r\lambda + \|z_\lambda\|)} = 0.$$

Therefore computing the limit in (3.90) we deduce that $\langle f(x_r), y \rangle \geq 0$ and we have that formula (3.89) is true.   □

Although not used in the sequel, it is worth noting the following result which is a kind of converse for Theorem 3.141.

PROPOSITION 3.12 *Let $(H, \langle \cdot, \cdot \rangle)$ be a Hilbert space, $K \subset H$ a closed convex cone, and $f : H \to H$ a mapping. If $\{x_r\}_{r>0}$ is a REFE of f with respect to K, then for any $r > 0$, x_r is a fixed point of the mapping $\Psi_r = P_{K_r} \circ (I - f)$.*

Proof. Suppose that x_r is a REFE. We have to prove that for any $r > 0$ $x_r = P_{K_r}(x_r - f(x_r))$, which is equivalent to

$$\langle f(x_r), x - x_r \rangle \geq 0, \tag{3.91}$$

for any $x \in K_r$, because the projection $P_K(x)$ of $x \in H$ onto a closed convex cone K is characterized by the relation $\langle x - P_K(x), y - P_K(x) \rangle \leq 0$, for any $y \in K$. Because $f(x_r) = u_r - \mu_r x_r$, relation (3.91) is equivalent to

$$\langle u_r - \mu_r x_r, x - x_r \rangle \geq 0. \tag{3.92}$$

Relation (3.92) is equivalent to

$$\langle u_r, x \rangle - \langle u_r, x_r \rangle - \mu_r \langle x_r, x \rangle + \mu_r \langle x_r, x_r \rangle \geq 0, \tag{3.93}$$

for any $x \in K_r$.

From the definition of u_r (u_r is in K^* and $\langle u_r, x_r \rangle = 0$), the second term of (3.93) is 0 and the first term is nonnegative. Because $\|x_r\| = r$ and μ_r is positive, from the Cauchy inequality applied to $\langle x_r, x \rangle$, the remaining part of (3.93) is also nonnegative. Hence (3.93) is true, proving that (3.91) is also true. $\square$

By Lemma 3.140 Theorem 3.141 has the following consequence.

COROLLARY 3.142 *Let $(H, \langle \cdot, \cdot \rangle)$ be a Hilbert space, $K \subset H$ a closed convex cone, and $f : H \to H$ a mapping such that for any $r > 0$ the mapping $\Psi_r(x) = P_{K_r}(x - f(x))$ has a fixed point (which is necessarily an element of K_r), where the mapping Ψ_r is considered from K_r into K_r. If f is without a REFE with respect to K, then the problem $\mathrm{NCP}(f, K)$ has a solution.*

Examples

We give several examples of REFE-acceptable mappings.

(I) In the n-dimensional Euclidean space $(\mathbb{R}^n, \langle \cdot, \cdot \rangle)$, any continuous mapping is REFE-acceptable with respect to any closed convex cone.

(II) Let $(H, \langle \cdot, \cdot \rangle)$ be an arbitrary Hilbert space and $K \subset H$ a closed convex cone with a compact base. It is known that in this case K is locally compact. Consequently, for any $r > 0$, K_r is a compact set. In this case, any continuous mapping f is REFE-acceptable with respect to K. This result is a consequence of Schauder's fixed point theorem.

(III) Let $(H, \langle \cdot, \cdot \rangle)$ be an arbitrary Hilbert space, $K \subset H$ an arbitrary closed convex cone, and $f : H \to H$ a completely continuous field; that is, f has a representation of the form $f(x) = x - T(x)$, where $T : H \to H$ is a completely continuous operator. In this case f is a REFE-acceptable mapping. This result is also a consequence of Schauder's fixed point theorem.

(IV) Let $(H, \langle \cdot, \cdot \rangle)$ be an arbitrary Hilbert space, $K \subset H$ an arbitrary closed convex cone, and $f : H \to H$ a nonexpansive field; that is, f has a representation of the form $f(x) = x - T(x)$, where $T : H \to H$ is a nonexpansive mapping. In this case, f is also a REFE-acceptable mapping. This result is a consequence of a classical fixed point theorem for nonexpansive mappings defined on a bounded closed convex subset of a uniformly convex Banach space.

(V) Let $(H, \langle \cdot, \cdot \rangle)$ be an arbitrary Hilbert space, $K \subset H$ an arbitrary closed convex cone, and $f : H \to H$ an α-set contraction field with respect to the α-Kuratowski measure of noncompactness. We have that $f(x) = x - T(x)$, where $T : H \to H$ is an α-set contraction. The mapping f is REFE-acceptable with respect to K. This result is a consequence of Darbo's fixed point theorem.

(VI) Let $(H, \langle \cdot, \cdot \rangle)$ be an arbitrary Hilbert space, and $K \subset H$ an arbitrary closed convex cone. Any mapping $f : H \to H$ with the property that for any $r > 0$, $VI(f, K_r)$ has a solution is REFE-acceptable with respect to K. An interesting example of such a mapping is a continuous quasi-monotone mapping $f : K \to H$. We recall that f is quasi-monotone on K if for any $x, y \in K$ the inequality $\langle f(x), y - x \rangle > 0$ implies $\langle f(y), y - x \rangle \geq 0$. Any pseudo-monotone mapping (in Karamardian's sense) is quasi-monotone. Also, in particular any monotone mapping is quasi-monotone. From Lemma 2.1 and Proposition 2.1, both proved in [Aussel and Hadjisavvas, 2004], we deduce that for any $r > 0$ the problem $VI(f, K_r)$ has a solution, because K_r is weakly compact. Because any solution of $VI(f, K_r)$ is a fixed point for the mapping Ψ_r, we have that any continuous quasi-monotone mapping is REFE-acceptable. About the solvability of the problem $VI(f, K_r)$ when f is quasi-monotone see also [Bianchi et al., 2004], Propositions 2.2 and 2.3.

With the next theorem we obtain other examples of REFE-acceptable mappings.

Let $(H, \langle \cdot, \cdot \rangle)$ be a Hilbert space. We recall the following notion defined by Isac in [Isac and Gowda, 1993]. Let D be a subset in H. We recall the definition of condition $(S)_+^1$.

DEFINITION 3.143 *We say that a mapping $f : D \to H$ satisfies condition $(S)^1_+$ if any sequence $\{x_n\}_{n \in \mathbb{N}} \subset D$ with (w)-$\lim_{n \to \infty} x_n = x_* \in H$, (w)-$\lim_{n \to \infty} f(x_n) = u \in H$ and $\limsup_{n \to \infty} \langle x_n, f(x_n) \rangle \leq \langle x_*, u \rangle$ has a subsequence $\{x_{n_k}\}_{k \in \mathbb{N}}$ convergent (in norm) to x_*, where (w)-lim denotes the weak limit.*

REMARK 3.14 *Condition $(S)^1_+$ is related to condition $(S)_+$ introduced in nonlinear analysis by Browder. It is known that condition $(S)_+$ implies condition $(S)^1_+$ [Isac, 2000d; Isac and Gowda, 1993]. Condition $(S)^1_+$ was used and considered in several papers (see the references in [Isac, 2000d]).*

We recall the following property of the inner product.

LEMMA 3.144 *If a sequence $\{x_n\}_{n \in \mathbb{N}}$ is weakly convergent to an element x_* and a sequence $\{y_n\}_{n \in \mathbb{N}}$ is convergent in norm to an element y_*, then $\lim_{n \to \infty} \langle x_n, y_n \rangle = \langle x_*, y_* \rangle$.*

DEFINITION 3.145 *We say that a mapping $f : H \to H$ is scalarly compact with respect to a closed convex set $D \subset H$, if for any sequence $\{x_n\}_{n \in \mathbb{N}} \subset D$ weakly convergent to an element $x_* \in D$ there exists a subsequence $\{x_{n_k}\}_{k \in \mathbb{N}}$ such that*

$$\limsup_{k \to \infty} \langle x_{n_k} - x_*, f(x_{n_k}) \rangle \leq 0.$$

REMARK 3.15 *If f is completely continuous or there exists a completely continuous operator $T : H \to H$ such that $|\langle y, f(x) \rangle| \leq \langle y, T(x) \rangle$ for any $x, y \in D$, then f is scalarly compact.*

A mapping $f : H \to H$ is called demicontinuous if for any sequence $\{x_n\}_{n \in \mathbb{N}} \subset H$ convergent in norm to an element $x_* \in H$, $\{f(x_n)\}_{n \in \mathbb{N}}$ is weakly convergent to $f(x_*)$.

THEOREM 3.146 *Let $(H, \langle \cdot, \cdot \rangle)$ be a Hilbert space and $T_1, T_2 : H \to H$ two demicontinuous mappings. If the following assumptions are satisfied.*

1. *T_1 is bounded and satisfies condition $(S)^1_+$;*

2. *T_2 is scalarly compact with respect to a closed bounded convex set $D \subset H$;*

then the problem $\mathrm{VI}(T_1 - T_2, D)$ has a solution.

Proof. Let Λ be the family of finite-dimensional subspaces F of H such that $F \cap D \neq \emptyset$. We consider the family Λ ordered by inclusion and also consider the mapping $h(x) = T_1(x) - T_2(x)$ defined for all $x \in D$. For each $F \in \Lambda$ we denote $D(F) = D \cap F$ and we set

$$A_F = \{y \in D \mid \langle x - y, h(y) \rangle \geq 0 \text{ for all } x \in D(F)\}.$$

For each $F \in \Lambda$ the set A_F is nonempty. Indeed, the solution set of $VI(h, D(F))$ is a subset of A_F and the solution set of $VI(h, D(F))$ is nonempty. To obtain this fact we consider the mappings $j : F \to H$, $j^* : H^* \to F^*$ and $j^* \circ h \circ j$, where j is the inclusion and j^* is the adjoint of j. The mappings $j^* \circ h \circ j$ are continuous and $\langle x - y, (j^* \circ h \circ j)(y) \rangle = \langle x - y, h(y) \rangle$. Applying the classical Hartman–Stampacchia theorem to the set $D(F)$ and to the mapping $j^* \circ h \circ j$ we obtain that $VI(h, D(h))$ has a solution. We denote by $\overline{A}_F^\sigma$ the weak closure of A_F. We have that $\cap_{F \in \Lambda} \overline{A}_F^\sigma$ is nonempty. Indeed, let $\overline{A}_{F_1}^\sigma, \overline{A}_{F_2}^\sigma, \ldots, \overline{A}_{F_n}^\sigma$ be a finite subfamily of the family $\{\overline{A}_F^\sigma\}_{F \in \Lambda}$. Let F_0 be the finite-dimensional subspace of H generated by $F_1, F_2, \ldots, F_n$. Because $F_k \subset F_0$ for all $k = 1, 2, \ldots, n$, we have that $D(F_k) \subset D(F_0)$, for all $k = 1, 2, \ldots, n$. We have $A_{F_0} \subset A_{F_K}$ which implies $\overline{A}_{F_0}^\sigma \subset \overline{A}_{F_K}^\sigma$, for all $k = 1, 2, \ldots, n$ and finally we deduce that $\cap_{k=1}^n \overline{A}_{F_k}^\sigma \neq \emptyset$. Because D is weakly compact we conclude that $\cap_{F \in \Lambda} \overline{A}_F^\sigma \neq \emptyset$. Let $y_* \in \cap_{F \in \Lambda} \overline{A}_F^\sigma$; that is, for any $F \in \Lambda$, $y_* \in \overline{A}_F^\sigma$. Let $x \in D$ be an arbitrary element. There exists some $F \in \Lambda$ such that $x, y_* \in F$. Because $y_* \in \overline{A}_F^\sigma$, by Smulian's theorem, there exists a sequence $\{y_n\}_{n \in \mathbb{N}}$, weakly convergent to y_*. We have

$$\langle y_* - y_n, h(y_n) \rangle \geq 0$$

and

$$\langle x - y_n, h(y_n) \rangle \geq 0$$

or

$$\langle y_n - y_*, T_1(y_n) \rangle \leq \langle y_n - y_*, T_2(y_n) \rangle \tag{3.94}$$

and

$$\langle x - y_n, T_1(y_n) \rangle \geq \langle x - y_n, T_2(y_n) \rangle. \tag{3.95}$$

From (3.94) and assumption 2 (considering eventually a subsequence of $\{y_{n \in \mathbb{N}}\}$), we have

$$\limsup_{n \to \infty} \langle y_n - y_*, T_1(y_n) \rangle \leq 0. \tag{3.96}$$

Because T_1 is bounded, we can suppose (taking eventually a subsequence of $\{y_{n \in \mathbb{N}}\}$) that $\{T_1(y_n)\}_{n \in \mathbb{N}}$ is weakly convergent to an element $v_0 \in H$. Because

$$\langle y_n, T_1(y_n) \rangle = \langle y_n - y_* + y_*, T_1(y_n) \rangle = \langle y_n - y_*, T_1(y_n) \rangle + \langle y_*, T_1(y_n) \rangle$$

and considering (3.96) we obtain

$$\limsup_{n \to \infty} \langle y_n, T_1(y_n) \rangle \leq \langle y_*, v_0 \rangle.$$

Hence, by condition $(S)^1_+$ we obtain that the sequence $\{y_n\}_{n \in \mathbb{N}}$ has a subsequence denoted again by $\{y_n\}_{n \in \mathbb{N}}$ convergent in norm to y_*. Because T_1 and T_2 are demicontinuous, we have (w)-$\lim_{n \to \infty} T_i(y_n) = T_i(y_*)$, for $i = 1, 2$. From inequality (3.95), by using Lemma 3.144 and computing the limit we conclude that

$$\langle x - y_*, T_1(y_*) - T_2(y_*) \rangle \geq 0,$$

for all $x \in D$, hence the proof is complete.   □

COROLLARY 3.147 *Let* $(H, \langle \cdot, \cdot \rangle)$ *be a Hilbert space,* $K \subset H$ *a closed convex cone, and* $f : H \to H$ *a mapping. If* f *has a decomposition of the form* $f(x) = T_1(x) - T_2(x)$ *such that*

1. T_1 *is demicontinuous, bounded, and satisfies condition* $(S)^1_+$;

2. T_2 *is demicontinuous and scalarly compact with respect to* K;

then f *is REFE-acceptable with respect to* K.

Proof. We apply Theorem 3.146 to f and any K_r with $r > 0$.   □

3.10.2   Mappings Without Regular Exceptional Family of Elements. A Necessary and Sufficient Condition

THEOREM 3.148 *Let* $(H, \langle \cdot, \cdot \rangle)$ *be a Hilbert space,* $K \subset H$ *a closed convex cone, and* $f : H \to H$ *a mapping. A necessary and sufficient condition for the mapping* f *to have the property of being without a REFE with respect to* K *is the following.*

There is a $\rho > 0$ *such that for any* $x \in K$ *with* $\|x\| = \rho$ *at least one of the following conditions holds.*

1. $\langle f(x), x \rangle \geq 0$.

2. *There is a* $y \in K$ *such that* $\rho^2 \langle f(x), y \rangle < \langle x, y \rangle \langle f(x), x \rangle$.

Proof. First suppose that f is without a REFE with respect to K and prove that at least one of the conditions given in the theorem is satisfied. Suppose to the contrary, that for any $r > 0$ there is an $x_r \in K$ with $\|x_r\| = r$ such that the following conditions hold.

(a) $\langle f(x_r), x_r \rangle < 0$.

(b) $r^2 \langle f(x_r), y \rangle \geq \langle x_r, y \rangle \langle f(x_r), x_r \rangle$, for any $y \in K$.

Let

$$\mu_r = -\frac{\langle f(x_r), x_r \rangle}{r^2}.$$

Then, by condition (a),

$$\mu_r > 0. \tag{3.97}$$

Let $u_r = \mu_r x_r + f(x_r)$. Then,

$$\langle u_r, x_r \rangle = 0. \tag{3.98}$$

Dividing condition (b) by r^2, we have

$$\langle f(x_r), y \rangle \geq -\mu_r \langle x_r, y \rangle,$$

for any $y \in K$. Hence,

$$\langle u_r, y \rangle \geq 0,$$

for any $y \in K$; that is,

$$u_r \in K^*. \tag{3.99}$$

Because $\|x_r\| = r$ for any $r > 0$, relations (3.97) through (3.99) imply that $\{x_r\}_{r>0} \subset K$ is a REFE for f with respect to K. But this is a contradiction. Hence, at least one of the conditions given in the theorem is satisfied.

Conversely, suppose that at least one of the conditions given in the theorem is satisfied and prove that f is without a REFE with respect to K. Suppose to the contrary that $\{x_r\}_{r>0} \subset K$ is a REFE for f with respect to K with corresponding μ_r and u_r (as given in Definition 3.138). Because $u_\rho = \mu_\rho x_\rho + f(x_\rho)$ and $\langle u_\rho, x_\rho \rangle = 0$, we have

$$0 < \mu_\rho = -\frac{\langle f(x_\rho), x_\rho \rangle}{\rho^2}.$$

Hence, $\langle f(x_\rho), x_\rho \rangle < 0$. Because $\|x_\rho\| = \rho$, the previous relation implies that for x_ρ condition 1 of the theorem is not satisfied. Hence, for x_ρ condition 2 of the theorem must hold; that is,

$$\rho^2 \langle f(x_\rho), y \rangle < \langle x_\rho, y \rangle \langle f(x_\rho), x_\rho \rangle, \tag{3.100}$$

for some $y \in K$. Dividing (3.100) by ρ^2 we obtain that

$$\langle f(x_\rho), y \rangle < -\mu_\rho \langle x_\rho, y \rangle,$$

and therefore

$$\langle u_\rho, y \rangle < 0.$$

Hence, $u_\rho \notin K^*$. But this contradicts condition 1. of Definition 3.138. Hence, f is without a REFE with respect to K.

By Lemma 3.140 and Theorem 3.148, we have as follows.

THEOREM 3.149 *Let $(H, \langle \cdot, \cdot \rangle)$ be a Hilbert space, $K \subset H$ a closed convex cone, and $f : H \to H$ a REFE-acceptable mapping. If there is a $\rho > 0$ such that for any $x \in K$ with $\|x\| = \rho$ at least one of the following conditions holds,*

1. $\langle f(x), x \rangle \geq 0$;

2. *there is an $y \in K$ such that $\rho^2 \langle f(x), y \rangle < \langle x, y \rangle \langle f(x), x \rangle$;*

then the problem $\mathrm{NCP}(f, K)$ has a solution.

THEOREM 3.150 *Let $(H, \langle \cdot, \cdot \rangle)$ be a Hilbert space, $K \subset H$ a closed convex cone, and $f : K \to H$ a REFE-acceptable mapping. If there is a mapping $h : K \to K$ with $h(0) = 0$ and such that at least one of the following conditions holds,*

1. $\underline{\mathcal{I}(f)^{\#}}(0) > 0$;

2. $\overline{(\mathcal{I}(f), h)}^{\#}(0) < \underline{h}^{\#}(0) \underline{\mathcal{I}(f)}^{\#}(0)$;

then the problem $\mathrm{NCP}(f, K)$ has a solution.

Proof. Suppose that condition 1. of Theorem 3.150 holds. Then, there is $\rho > 0$ such that $\langle \mathcal{I}(f)(u), u \rangle \geq 0$, for any $u \in K \backslash \{0\}$ with $\|u\| \leq \rho$. Let $x = i(u)$. Then $x \in K$, $\|x\| \geq \rho$, and

$$0 \leq \|u\|^2 \langle f(i(u)), u \rangle = \frac{1}{\|x\|^4} \langle f(x), x \rangle.$$

Hence, $\langle f(x), x \rangle \geq 0$, for any $x \in K$ with $\|x\| \geq \rho$. Thus, condition 1. of Theorem 3.149 holds, proving that the problem $\mathrm{NCP}(f, K)$ has a solution.

Suppose now that condition 2 of Theorem 3.150 holds. Then,

$$\limsup_{\substack{u \to 0 \\ u \in K}} \frac{\langle \mathcal{I}(f)(u), h(u) \rangle}{\|u\|^2} < \liminf_{\substack{u \to 0 \\ u \in K}} \frac{\langle u, h(u) \rangle}{\|u\|^2} \liminf_{\substack{u \to 0 \\ u \in K}} \frac{\langle \mathcal{I}(f)(u), u \rangle}{\|u\|^2}$$

$$\leq \liminf_{\substack{u \to 0 \\ u \in K}} \left(\frac{\langle u, h(u) \rangle}{\|u\|^2} \frac{\langle \mathcal{I}(f)(u), u \rangle}{\|u\|^2} \right).$$

Hence,

$$\limsup_{\substack{u \to 0 \\ u \in K}} \left(\langle f(i(u)), h(u) \rangle - \frac{\langle u, h(u) \rangle}{\|u\|^2} \langle f(i(u)), u \rangle \right) < 0.$$

With the nonlinear coordinate transformation $u = i(x)$ (i.e., $x = i(u)$) we have

$$\limsup_{\substack{\|x\| \to +\infty \\ x \in K}} \left(\langle f(x), h(i(x)) \rangle - \langle x, h(i(x)) \rangle \frac{\langle f(x), x \rangle}{\|x\|^2} \right) < 0.$$

Hence, there is a $\rho > 0$ such that for any $x \in K$ with $\|x\| \geq \rho$ we have

$$\langle f(x), h(i(x)) \rangle - \langle x, h(i(x)) \rangle \frac{\langle f(x), x \rangle}{\|x\|^2} < 0. \qquad (3.101)$$

Let $y = h(i(x)) \in K$. Then, Equation (3.101) implies that condition 2 of Theorem 3.149 is satisfied. Hence, the problem $\mathrm{NCP}(f, K)$ has a solution. $\square$

REMARK 3.16 *Theorem 3.148 can be used to find new existence theorems for complementarity problems. This subject must be developed.*

Chapter 4

Scalar Derivatives in Banach Spaces

4.1 Preliminaries

Let E be a Banach space and E^* the topological dual of E. Let $\langle E, E^* \rangle$ be a duality between E and E^*. This duality is with respect to a bilinear functional on $E \times E^*$ denoted $\langle \cdot, \cdot \rangle$ and which satisfies the following separation axioms:

(s_1): $\langle x_0, y \rangle = 0$ for all $y \in E^*$ implies $x_0 = 0$,

(s_2): $\langle x, y_0 \rangle = 0$ for all $x \in E$ implies $y_0 = 0$.

For the weak topology on E (resp., on E^*) we use Bourbaki's terminology; that is, the weak topology on E is the $\sigma(E, E^*)$-topology and on E^* the $\sigma(E^*, E)$-topology. Denote by $L(E, E^*)$ the set of continuous linear mappings from E into E^*. We remark that if $E = H$, where H is a Hilbert space, then E^* can be identified with H, the bilinear functional generating the duality between E and E^* with the scalar product of H and $L(E, E^*)$ with the space of continuous linear mappings from H into H, which are denoted $L(H)$ [Kantorovici and Akilov, 1977].

Recall the following definitions [Isac, 2000d].

DEFINITION 4.1 *Let $K \subseteq E$ and $f : K \rightarrow E^*$. f is called* completely continuous *if it is continuous and the image of every bounded set is relatively compact.*

DEFINITION 4.2 *We say that a nonempty set $K \subseteq E$ is a convex cone if:*

1. $K + K \subseteq K$.

2. $\lambda K \subseteq K$ for all $\lambda \in \mathbb{R}_+$.

A convex cone K is called pointed if $K \cap (-K) = \{0\}$ and generating if $K - K = E$.

DEFINITION 4.3 *Let $K \subseteq E$ be a convex cone. The convex cone*

$$K^* = \{y \in E^* \mid \langle x, y \rangle \geq 0 \text{ for all } x \in K\}$$

of E^* is called the dual cone of K.

For more details about convex cones the reader is referred to [Isac, 2000d].

DEFINITION 4.4 *Let D be a set, $K \subseteq E$ a pointed convex cone, $x, y \in K$, and $f, g : D \to E$. The relation $x \leq_K y$ defined by $y - x \in K$ is an order, relation on E. Define $f \leq_K g$ if $f(z) \leq_K g(z)$ for all $z \in D$.*

Let $(H, \langle \cdot, \cdot \rangle)$ be a Hilbert space. Recall the following definitions.

DEFINITION 4.5 *A continuous operator $Z : H \to H$ is called skew-adjoint* [Atiyah and Singer, 1969] *if*

$$\langle Z(x), y \rangle = -\langle Z(y), x \rangle \tag{4.1}$$

for all $x, y \in H$. In [Nemeth, 1992] *it is proved that relation (4.1) implies that Z is linear.*

DEFINITION 4.6 *A continuous linear operator $P : H \to H$ is called positive semi-definite* [Riesz and Nagy, 1990] *if*

$$\langle P(x), x \rangle \geq 0,$$

for all $x \in H$.

4.2   Semi-inner Products

Let $(E, \|\cdot\|)$ be an arbitrary real Banach space. We say that a *semi-inner product* (in Lumer's sense) is defined on E, if to any $x, y \in E$ there corresponds a real number denoted $[x, y]$ satisfying the following properties.

(s_1) $[x + y, z] = [x, z] + [y, z]$.

(s_2) $[\lambda x, y] = \lambda [x, y]$, for $x, y, z \in E, \lambda \in \mathbb{R}$.

(s_3) $[x, x] > 0$ for $x \neq 0$.

(s_4) $|[x, y]|^2 \leq [x, x][y, y]$.

It is known [Giles, 1967; Lumer, 1961] that a semi-inner product space is a normed linear space with the norm $\|x\|_s = [x, x]^{1/2}$ and that every Banach space can be endowed with a semi-inner product (and in general in infinitely many different ways, but a Hilbert space in a unique way).

Obviously if $(H, \langle \cdot, \cdot \rangle)$ is a Hilbert space, the inner product $\langle \cdot, \cdot \rangle$ is the unique semi-inner product in Lumer's sense on H [Giles, 1967; Lumer, 1961]. We note that it is possible to define a semi-inner product such that $[x, x] = \|x\|^2$ (where $\| \cdot \|$ is the norm given in E). In this case we say that the semi-inner product is *compatible with the norm* $\| \cdot \|$. By the proof of Theorem 1 of [Giles, 1967] this semi-inner product can be defined so that it has the homogeneity property:

(s_5) $[x, \lambda y] = \lambda [x, y]$, for $x, y \in E$, $\lambda \in \mathbb{R}$.

Throughout this chapter we suppose that all semi-inner products compatible with the norm satisfy (s_5).

4.3 Inversions

We recall again the following definition which is an extension of [do Carmo, 1992], Example 5.1, p. 169.

DEFINITION 4.7 *The operator*

$$i : E\backslash\{0\} \to E\backslash\{0\}; \; i(x) = \frac{x}{[x, x]}$$

is called the inversion (of pole 0) *with respect to* $[\cdot, \cdot]$.

It is easy to see that i is one-to-one and $i^{-1} = i$. Indeed, because

$$\|i(x)\|_s = \frac{1}{\|x\|_s},$$

by the definition of i we have

$$i(i(x)) = \frac{i(x)}{\|i(x)\|_s^2} = \|x\|_s^2 i(x) = x.$$

Hence i is a global homeomorphism of $E\backslash\{0\}$ which can be viewed as a global nonlinear coordinate transformation in E.

Let $A \subseteq E$ such that $0 \in A$ and $A\backslash\{0\}$ is an invariant set of the inversion i with respect to $[\cdot, \cdot]$; that is, $i(A\backslash\{0\}) = A\backslash\{0\}$ and $f : A \to E$. Examples of invariant sets of the inversion i with respect to $[\cdot, \cdot]$ are:

1. $F\backslash\{0\}$ where F is a linear subspace of E (in particular F can be the whole E).

2. $K\backslash\{0\}$ where $K \subseteq E$ is a convex cone.

Now we define the inversion (of pole 0) with respect to $[\cdot, \cdot]$ of the mapping f.

DEFINITION 4.8 *The* inversion (of pole 0) *with respect to* $[\cdot, \cdot]$ *of the mapping* f *is the mapping* $\mathcal{I}(f) : A \to E$ *defined by:*

$$\mathcal{I}(f)(x) = \begin{cases} [x, x](f \circ i)(x) & \text{if} \quad x \neq 0, \\ 0 & \text{if} \quad x = 0. \end{cases}$$

PROPOSITION 4.1 *The inversion of mappings* $\mathcal{I}$ *with respect to* $[\cdot, \cdot]$ *is a one-to-one operator on the set of mappings* $\{f \mid f : A \to E; f(0) = 0\}$ *and* $\mathcal{I}^{-1} = \mathcal{I}$; *that is,* $\mathcal{I}(\mathcal{I}(f)) = f$.

Proof. By definition $\mathcal{I}(\mathcal{I}(f))(0) = 0$. Hence, $\mathcal{I}(\mathcal{I}(f))(0) = f(0)$. If $x \neq 0$, then $\mathcal{I}(\mathcal{I}(f))(x) = \|x\|_s^2 \mathcal{I}(f)(i(x)) = \|x\|_s^2 \|i(x)\|_s^2 f(i(i(x))) = f(x)$. Thus, $\mathcal{I}(\mathcal{I}(f))(x) = f(x)$ for all $x \in A$. Therefore, $\mathcal{I}(\mathcal{I}(f)) = f$. $\square$

PROPOSITION 4.2 *Let* $f : A \to A$. *Then,* $x \neq 0$ *is a fixed point of* f *iff* $i(x)$ *is a fixed point of* $\mathcal{I}(f)$.

Proof. Suppose that $x \neq 0$ is a fixed point of f; that is, $f(x) = x$. Because $i(i(x)) = x$ we have

$$f(i(i(x))) = x. \tag{4.2}$$

Multiplying (4.2) by

$$\|i(x)\|_s^2 = \frac{1}{\|x\|_s^2}$$

we obtain $\mathcal{I}(f)(i(x)) = i(x)$. Thus, $i(x)$ is a fixed point of $\mathcal{I}(f)$. Similarly it can be proved that if $i(x)$ is a fixed point of $\mathcal{I}(f)$, then x is a fixed point of f. $\square$

Let $D = \{x \in E \mid \|x\|_s \leq 1\}$ and $C = \{x \in E \mid \|x\|_s = 1\}$.

PROPOSITION 4.3 *Let* $f, g : A \to E$ *such that* $f(x) = g(x)$ *for all* $x \in A \cap C$ *and* $f(0) = g(0) = 0$. *There exist unique extensions* $\tilde{f}, \tilde{g} : A \to E$ *of* $f|_{A \cap D}$ *and* $g|_{A \cap D}$, *respectively, such that* $\tilde{g} = \mathcal{I}(\tilde{f})$.

Proof. Let $D^\circ = \{x \in E \mid \|x\|_s < 1\}$. First we prove the existence of the extensions $\tilde{f}, \tilde{g}$. Define the extensions $\tilde{f}, \tilde{g}$ of $f|_{A \cap D}$ and $g|_{A \cap D}$ by

$$\tilde{g}(x) = \begin{cases} g(x), & \text{if } \|x\|_s \leq 1 \\ \mathcal{I}(f)(x), & \text{if } \|x\|_s > 1 \end{cases}$$

and

$$\tilde{f}(x) = \begin{cases} f(x), & \text{if } \|x\|_s \leq 1 \\ \mathcal{I}(g)(x), & \text{if } \|x\|_s > 1, \end{cases}$$

respectively. We have to prove that

$$\tilde{g}(x) = \mathcal{I}(\tilde{f})(x) \tag{4.3}$$

for all $x \in A$. We consider three cases

First case: $x \in A \cap D^\circ$. In this case $\|x\|_s < 1$ and hence, $\|i(x)\|_s > 1$. Thus, by definition $\tilde{g}(x) = g(x)$ and $\tilde{f}(i(x)) = \mathcal{I}(g)(i(x))$. By using these relations and the definition of the inversion of a mapping with respect to a semi-inner product, relation (4.3) can be proved easily.

Second case: $x \in A \backslash D$. In this case $\|x\|_s > 1$ and hence, $\|i(x)\|_s < 1$. Thus, by definition $\tilde{g}(x) = \mathcal{I}(f)(x)$ and $\tilde{f}(i(x)) = f(i(x))$. Relation (4.3) can be proved similarly to the previous case.

Third case: $x \in A \cap C$. In this case $\|x\|_s = 1$ and hence, $i(x) = x$. Thus, by definition $\tilde{g}(x) = g(x)$ and $\tilde{f}(i(x)) = f(x)$. In this case (4.3) is equivalent to $f(x) = g(x)$, which by the assumption made on f and g it is true.

Now we prove the uniqueness of the extensions $\tilde{f}, \tilde{g}$. Suppose that $\hat{f}, \hat{g}$ are extensions of $f|_{A \cap D}$ and $g|_{A \cap D}$, respectively, such that $\hat{g} = \mathcal{I}(\hat{f})$. If $\|x\|_s \leq 1$, then $\hat{g}(x) = \tilde{g}(x) = g(x)$ becauce both $\hat{g}$ and $\tilde{g}$ are extensions of $g|_{A \cap D}$. If $\|x\|_s > 1$, then $\|i(x)\|_s < 1$. Because $\hat{f}$ is an extension of $f|_{A \cap D}$, $\hat{f}(i(x)) = f(i(x))$. By using this relation, relation $\hat{g}(x) = \mathcal{I}(\hat{f})(x)$, the definition of the inversion of a mapping with respect to a semi-inner product and the definition of $\tilde{g}$ we obtain $\hat{g}(x) = \tilde{g}(x)$. Hence, $\hat{g} = \tilde{g}$. Relation $\hat{g} = \mathcal{I}(\hat{f})$ implies $\hat{f} = \mathcal{I}(\hat{g})$. Hence relation $\hat{f} = \tilde{f}$ can be proved by interchanging the roles of f and g. $\square$

In the case of $f = g$ Proposition 4.3 has the following corollary.

COROLLARY 4.9 *Let* $f : A \to E$; $f(0) = 0$. *There exists a unique extension* $\tilde{f} : A \to E$ *of* $f|_{A \cap D}$ *such that* $\tilde{f}$ *is a fixed point of* $\mathcal{I}$ *(i.e.,* $\tilde{f} = \mathcal{I}(\tilde{f})$*).*

REMARK 4.1 *It is easy to see that the inversion of mappings with respect to* $[\cdot, \cdot]$ *is linear and has the following properties.*

1. *If* $T \in L(E, E)$ *and* $j : A \hookrightarrow E$ *is the embedding of* A *into* E, *then* $\mathcal{I}(T \circ j) = T \circ j$.

2. *If the semi-inner product is compatible with the norm of* E *and* $\|x\| \to +\infty$, *then* $i(x) \to 0$.

4.4 Scalar Derivatives

Let $(E, \| \cdot \|)$ be an arbitrary real Banach space and $[\cdot, \cdot]$ a semi-inner product on E. Let $G \subseteq E$ be a set which contains at least one non isolated point, $\widetilde{G} \subseteq E$

such that $G \subseteq \widetilde{G}$, $f : \widetilde{G} \to E$, and x_0 a non isolated point of G. The following definition is an extension of Definition 2.2 of [Nemeth, 1992].

DEFINITION 4.10 *The limit*

$$\underline{f}^{\#,G}(x_0) = \liminf_{\substack{x \to x_0 \\ x \in G}} \frac{[f(x) - f(x_0), x - x_0]}{\|x - x_0\|_s^2}$$

is called the lower scalar derivative *of f at x_0 along G with respect to $[\cdot, \cdot]$. Taking* lim sup *in place of* lim inf, *we can define the* upper scalar derivative $\overline{f}^{\#,G}(x_0)$ *of f at x_0 along G with respect to $[\cdot, \cdot]$ similarly. If $G = \widetilde{G}$, then without confusion, we can say lower scalar derivative and upper scalar derivative, for short, instead of lower scalar derivative along G and upper scalar derivative along G, respectively. In this case, we omit G from the superscript of the corresponding notations.*

We have as follows.

LEMMA 4.11 *Suppose that $[\cdot, \cdot]$ is compatible with the norm $\| \cdot \|$. Let $K \subseteq E$ be an unbounded set such that $0 \in K$ and $K \backslash \{0\}$ is an invariant set of the inversion i with respect to $[\cdot, \cdot]$. Let $g : E \to E$. Then we have*

$$\liminf_{\substack{\|x\| \to \infty \\ x \in K}} \frac{[g(x), x]}{\|x\|^2} = \mathcal{I}(g)^{\#,K}(0).$$

Proof. Because $K \subseteq E$ is unbounded and $K \backslash \{0\}$ is an invariant set of i, 0 is a non-isolated point of K. Hence, $\mathcal{I}(g)^{\#,K}(0)$ is well defined. Consider the global nonlinear coordinate transformation $y = i(x)$. Then $x = i(y)$ and we have

$$\liminf_{\substack{\|x\| \to \infty \\ x \in K}} \frac{[g(x), x]}{\|x\|^2} = \liminf_{\substack{y \to 0 \\ y \in K}} [\mathcal{I}(g)(y), i(y)],$$

from where, by using the definition of the lower scalar derivative along a set, the assertion of the lemma follows easily. $\qquad \square$

4.5 Fixed Point Theorems in Banach Spaces

4.5.1 A Fixed Point Index for α-condensing Mappings

Let $(E, \|.\|)$ be a Banach space. For a bounded set D in E we denote by $\alpha(D)$ the *measure of noncompactness* of D defined by

$$\alpha(D) = \inf\{r > 0 \,|\, D \text{ admits a finite cover by sets of diameter at most } r\}.$$

For properties of $\alpha(D)$ see [Akhmerov et al., 1992; Banas and Goebel, 1980], and [Sadovskii, 1972].

A continuous mapping $f : \text{dom}(f) \subset E \to E$ is called *k-α-contractive* if there is a $k \geq 0$ such that $\alpha(f(D)) \leq k\alpha(D)$ for each bounded set $D \subset \text{dom}(f)$.

Also, f is called *α-condensing* if $\alpha(f(D)) < \alpha(D)$, for each bounded set $D \subset \text{dom}(f)$, with $\alpha(D) \neq 0$. We recall that f is *completely continuous* if f is continuous and for every bounded set $D \subset \text{dom}(f)$, we have $f(D)$ is relatively compact (i.e., $\overline{f(D)}$ is compact).

It is known that a completely continuous mapping is 0-α-contractive. Every k-α-contractive mapping with $0 \leq k < 1$ is α-condensing, but there are α-condensing mappings that are not k-α-contractive for any $k < 1$ [Akhmerov et al., 1992; Nussbaum, 1971; Sadovskii, 1972].

Let $K \subset E$ be a closed convex cone and let D be a bounded open set in E. Suppose that $D_k = D \cap K \neq \emptyset$. Denote by $\overline{D}_K$ the closure and ∂D_K the boundary of D_K relative to K.

We need to recall some properties of the measure of noncompactness α.

(α_1) $\alpha(A) = 0$ if and only if $\overline{A}$ is compact.

(α_2) $\alpha(A) = \alpha(\overline{A})$.

(α_3) $A_1 \subseteq A_2$ implies $\alpha(A_1) \leq \alpha(A_2)$.

(α_4) $\alpha(A \cup B) = \max\{\alpha(A), \alpha(B)\}$.

(α_5) $\alpha(\lambda A) = |\lambda|\alpha(A), \lambda \in \mathbb{R}$.

(α_6) $\alpha(\text{conv}(A)) = \alpha(A)$.

(α_7) $\alpha(A + B) \leq \alpha(A) + \alpha(B)$.

When $f : \overline{D}_K \to K$ is α-condensing and $f(x) \neq x$ for any $x \in \partial D_K$, there is defined in [Nussbaum, 1971] and [Sadovskii, 1972] an integer $i_K(f, D_K)$, called the *fixed point index* of f on D_K, which has the following properties.

(i_1) *(Existence property)*: if $i_K(f, D_K) \neq 0$, then f has a fixed point in D_K.

(i_2) *(Normalization)*: if $u \in D_K$, then $i_K(\hat{u}, D_K) = 1$, where $\hat{u}(x) = u$, for any $x \in \overline{D}_K$.

(i_3) *(Additivity property)*: if U_1, U_2 are disjoint relatively open subsets of D_K such that $f(x) \neq x$ for any $x \in \overline{D}_K \backslash (U_1 \cup U_2)$, then $i_K(f, D_K) = i_K(f, U_1) + i_K(f, U_2)$.

(i_4) *(Homotopy property)*: if $H : [0, 1] \times \overline{D}_K \to K$ is continuous and such that $\alpha(H([0, 1] \times A)) < \alpha(A)$, for each $A \subset D_K$ with $\alpha(A) \neq 0$ and if $H(t, x) \neq x$ for any $x \in \partial D_K$ and any $t \in [0, 1]$, then $i_K(H(0, \cdot), D_K) = i_K(H(1, \cdot), D_K)$.

4.5.2   An Altman-type Fixed Point Theorem

Let $(E, \| \cdot \|)$ be a Banach space and $K \subset E$ a closed convex cone. Suppose given a mapping $B : E \times E \to \mathbb{R}$ satisfying the following properties.

(b_1) $B(\lambda x, y) = \lambda B(x, y)$, for any $\lambda > 0$ and any $x, y \in E$.

(b_2) $B(x, x) > 0$, for any $x \in E$, $x \neq 0$.

EXAMPLE 4.12

1. *If E is a Hilbert space and $\langle \cdot, \cdot \rangle$ is the inner product on E, then $B(x, y) :=$ $\langle x, y \rangle$, for all $x, y \in E$.*

2. *Let $(E, \| \cdot \|)$ be an arbitrary Banach space and let $[\cdot, \cdot]$ be a semi-inner product as defined by Lumer [1961] and studied by Giles in [Giles, 1967]. In this case we take $B(x, y) = [x, y]$ for all $x, y \in E$. It is known [Giles, 1967; Lumer, 1961] that on any Banach space we can define a semi-inner product.*

3. *The real Banach space $\mathcal{L}_p(X, \mathcal{S}, \nu)$, where $1 < p \leq 2$, can be expressed as a uniform semi-inner product space with the semi-inner product*

$$[x, y] = \frac{1}{\|y\|_p^{p-2}} \int_X x |y|^{p-1} \mathrm{sgn}\, x \, d\nu,$$

for all $x, y \in \mathcal{L}_p(X, \mathcal{S}, \nu)$. In this case we take $B(x, y) = [x, y]$, for all $x, y \in \mathcal{L}_p(X, \mathcal{S}, \nu)$, compatible with the norm.

4. *Also in any Banach space we can consider*

$$B(x, y) = \|y\| \lim_{t \to 0_+} \frac{\|y + tx\| - \|y\|}{t}.$$

5. *Any coercive bilinear form $B : E \times E \to \mathbb{R}$ can also be used.*

THEOREM 4.13 *If $f : E \to E$ is α-condensing $f(K) \subseteq K$ and*

$$\limsup_{\substack{\|x\| \to \infty \\ x \in K}} \frac{B(f(x), x)}{B(x, x)} < 1,$$

then f has a fixed point in K.

Proof. Consider the continuous mapping $H : [0, 1] \times E \to E$ defined by $H(t, x) = tf(x)$. We show that there is an $R > 0$ sufficiently large such that for any $x \in K$ with $\|x\| = R$, and any $t \in [0, 1]$ we have $H(t, x) \neq x$. Indeed, if we suppose the contrary, then for every positive integer n there exists an

$x_n \in E$ and a $t_n \in [0, 1]$ such that $\|x_n\| = n$ and $H(t_n, x_n) = x_n$; that is, $t_n f(x_n) = x_n$. It follows that $t_n \neq 0$ and consequently $f(x_n) = t_n^{-1} x_n$, where $t_n^{-1} \geq 1$. Thus we have

$$\frac{B(f(x_n), x_n)}{B(x_n, x_n)} = t_n^{-1},$$

for all $n \in \mathbb{N}$. Because $\{t_n\}_{n \in \mathbb{N}} \subset [0, 1]$, there exists a convergent subsequence $\{t_{n_k}\}_{k \in \mathbb{N}}$ of $\{t_n\}_{n \in \mathbb{N}}$ such that $\lim_{k \to \infty} t_{n_k} = t^* \in [0, 1]$. The limit can be 0 or not. Hence in both situations we have that

$$\lim_{k \to \infty} \frac{B(f(x_{n_k}), x_{n_k})}{B(x_{n_k}, x_{n_k})}$$

exists and it is in $[1, +\infty]$. Because $\lim_{k \to \infty} \|x_{n_k}\| = +\infty$, we have that

$$\limsup_{\substack{\|x\| \to \infty \\ x \in K}} \frac{B(f(x), x)}{B(x, x)} \geq 1,$$

which is a contradiction. Hence there exists $R > 0$ with the property indicated above. Let $D = \{x \in E \mid \|x\| < R\}$ and $D_K = D \cap K$. Obviously $D_K \neq \emptyset$, because $0 \in D_K$ and we have that $\partial D_K = \{x \in K \mid \|x\| = R\}$. We denote again by H the restriction of H to the set $[0, 1] \times \overline{D}_K$. We have that H is a continuous homotopy and $H : [0, 1] \times \overline{D}_K \to K$. We have $H(t, x) \neq x$ for any $x \in \partial D_K$ and any $t \in [0, 1]$. Now we show that $\alpha(H([0, 1] \times A)) < \alpha(A)$ for each $A \subset D$ with $\alpha(A) \neq 0$. Inasmuch as

$$H([0, 1] \times A) = \bigcup_{0 \leq t \leq 1} t f(A)$$

and

$$\bigcup_{0 \leq t \leq 1} t f(A) \subseteq \text{conv}[f(A) \cup \{0\}],$$

then by applying the properties (α_1)–(α_6) of the measure of noncompactness α, we have

$$\alpha(H([0, 1] \times A)) = \alpha\left(\bigcup_{0 \leq t \leq 1} t f(A) \right) \leq \alpha(f(A)) < \alpha(A).$$

The assumption of property (i_4) of the fixed point index $i_k(f, D_K)$ is satisfied and we deduce that

$$i_K(H(0, \cdot), D_K) = i_K(H(1, \cdot), D_K).$$

Because $H(0, \cdot) : \overline{D}_K \to K$ is the mapping $H(0, x) = 0 \cdot f(x) = 0$ for any $x \in \overline{D}_K$ and $0 \in D_K$, we have by property (i_2) that $i_K(H(0, \cdot), D_K) = 1$, and therefore $i_K(H(1, \cdot), D_K) = 1$. Now, by property (i_1) we have that f has a fixed point in D_K; that is, in K. $\square$

COROLLARY 4.14 *If $[\cdot, \cdot]$ is a semi-inner product in E, then any α-condensing mapping $f : E \to E$ with $f(K) \subseteq K$ which has λI as an asymptotic scalar derivative along K with respect to $[\cdot, \cdot]$, with $0 < \lambda < 1$, has a fixed point in K.*

Proof. The proof is a straightforward consequence of Theorem 4.13 with $B = [\cdot, \cdot]$ and of the definition of the asymptotic scalar derivative along K with respect to $[\cdot, \cdot]$. $\square$

REMARK 4.2 *Theorem 4.13 has as a particular case Theorem 3.2.4, given in [Akhmerov et al., 1992], p. 106.*

THEOREM 4.15 *If $f : E \to E$ is α-condensing $f(K) \subseteq K$, $(I - f)(K) \subseteq K$, $[\cdot, \cdot]$ is a semi-inner product compatible with the norm $\| \cdot \|$ and*

$$\limsup_{\substack{\|x\| \to \infty \\ x \in K}} \frac{[f(x), x]}{[x, x]} < 1,$$

then $(I - f)|_K : K \to K$ is surjective.

Proof. By the condition (s_4) of the semi-inner product $[\cdot, \cdot]$ it follows that for all $y \in K$ the operator $f_y : K \to K$; $f_y(x) = f(x) + y$ satisfies the condition of Theorem 4.13 with $B = [\cdot, \cdot]$. Hence it has a fixed point; that is, $(I - f)|_K$ is surjective. $\square$

COROLLARY 4.16 *If $[\cdot, \cdot]$ is a semi-inner product in E compatible with the norm $\| \cdot \|$, then for any α-condensing mapping $f : E \to E$ with $f(K) \subseteq K$ and $(I - f)(K) \subseteq K$ which has λI as an asymptotic scalar derivative along K with respect to $[\cdot, \cdot]$, with $0 < \lambda < 1$, $(I - f)|_K : K \to K$ is surjective.*

Proof. The proof is a straightforward consequence of Theorem 4.15 and the definition of the asymptotic scalar derivative along K with respect to $[\cdot, \cdot]$. $\square$

THEOREM 4.17 *If $[\cdot, \cdot]$ is a semi-inner product in E compatible with the norm $\| \cdot \|$ and $f : E \to E$ is an α-condensing mapping with $f(K) \subseteq K$ and $\overline{\mathcal{I}(f)}^{\#, K}(0) < 1$, then f has a fixed point in K.*

Proof. Consider the global nonlinear coordinate transformation $y = i(x)$, where i is the inversion with respect to $[\cdot, \cdot]$. Then,

$$\overline{\mathcal{I}(f)}^{\#, K}(0) = \limsup_{\substack{\|x\| \to \infty \\ x \in K}} \frac{[f(x), x]}{\|x\|^2}$$

and the proof follows by applying Theorem 4.13. $\square$

THEOREM 4.18 *If* $[\cdot, \cdot]$ *is a semi-inner product in* E *compatible with the norm* $\| \cdot \|$ *and* $f : E \to E$ *is an* α-*condensing mapping with* $f(K) \subseteq K$, $(I - f)(K) \subseteq K$, *and* $\overline{\mathcal{I}(f)}^{\#,K}(0) < 1$, *then* $(I - f)|_K : K \to K$ *is surjective.*

Proof. Consider the global nonlinear coordinate transformation $y = i(x)$, where i is the inversion with respect to $[\cdot, \cdot]$. Then,

$$\overline{\mathcal{I}(f)}^{\#,K}(0) = \limsup_{\substack{\|x\| \to \infty \\ x \in K}} \frac{[f(x), x]}{\|x\|^2}$$

and the proof follows by applying Theorem 4.15. $\qquad\qquad\qquad\square$

4.5.3 Integral Equations

Let $\Omega \subseteq \mathbb{R}$ be a bounded open set, $L^2(\Omega)$ the set of functions on Ω whose square is integrable on Ω, and

$$L^2_+(\Omega) = \{u \in L^2(\Omega) \mid u(t) \geq 0 \text{ for almost all } t \in \Omega\}.$$

$L^2(\Omega)$ is a Hilbert space with respect to the scalar product

$$\langle u, v \rangle = \int_\Omega u(s)v(s)ds,$$

and $L^2_+(\Omega)$ is a generating closed convex pointed cone of $L^2(\Omega)$. Let $\mathcal{L} : \overline{\Omega} \times \overline{\Omega} \times \mathbb{R} \to \mathbb{R}$, $\mathcal{K} : \overline{\Omega} \times \overline{\Omega} \to \mathbb{R}$, and $\mathcal{F} : \overline{\Omega} \times \mathbb{R} \to \mathbb{R}$. Denote by $\mathcal{I}_3$ and $\mathcal{I}_2$ the inversions with respect to the third and second variable (considered as functions from Ω to $\mathbb{R}$), respectively, and by $[-\varepsilon, \varepsilon]^\Omega$ the set of functions from Ω to $\mathbb{R}$ with values in the interval $[-\varepsilon, \varepsilon]$. We recall the following definition and result [Zabreiko et al., 1975].

DEFINITION 4.19 *We say that* $\mathcal{L}$ *is a* Caratheodory *function if* $\mathcal{L}(s, t, u)$ *is continuous with respect to* u *for almost all* $(s, t) \in \overline{\Omega} \times \overline{\Omega}$ *and is measurable in* (s, t) *for each* $u \in \mathbb{R}$.

THEOREM 4.20 *If the following conditions are satisfied,*

1. $\mathcal{L}$ *is a Caratheodory function;*

2. $|\mathcal{L}(s, t, u)| \leq \mathcal{R}(s, t)(a + b|u|)$ *for almost all* $s, t \in \Omega$, $\forall u \in \mathbb{R}$, *where* $a, b > 0$ *and* $\mathcal{R} \in L^2(\Omega \times \Omega)$;

3. *For any* $\alpha > 0$ *the function* $\mathcal{R}_\alpha(s, t) = \max_{|u| \leq \alpha} |\mathcal{L}(s, t, u)|$ *is summable with respect to* t *for almost all* $s \in \Omega$;

4. *For any* $\alpha > 0$,

$$\lim_{\text{mes}(D)\to 0} \sup_{|u|\leq\alpha} \left\| \mathcal{P}_D \int_\Omega \mathcal{L}(s,t,u(t))dt \right\|_{L^2(\Omega)} = 0,$$

where $\text{mes}(D)$ *is the Lebesgue measure of* D *and* $\mathcal{P}_D$ *is the operator of multiplication by the characteristic function of the set* $D \subseteq \Omega$;

5. *For any* $\beta > 0$,

$$\lim_{\text{mes}(D)\to 0} \sup_{\|u\|_{L^2(\Omega)}\leq\beta} \left\| \int_\Omega \mathcal{L}(s,t,u(t))dt \right\|_{L^2(\Omega)} = 0;$$

then the operator

$$A(u)(s) = \int_\Omega \mathcal{L}(s,t,u(t))dt$$

is a completely continuous operator from $L^2(\Omega)$ *into* $L^2(\Omega)$.

The integral of an almost everywhere nonnegative function is nonnegative, therefore by Theorem 4.20 we have the following:

COROLLARY 4.21 *If conditions 1–5 of Theorem 4.20 and condition*

6. $\mathcal{L}(s,t,u) \geq 0$ *for all* $u \in \mathbb{R} \cap [0,+\infty[$, *for all* $s \in \Omega$, *and for almost all* $t \in \Omega$

are satisfied, then the operator

$$A(u)(s) = \int_\Omega \mathcal{L}(s,t,u(t))dt$$

is a completely continuous operator from $L^2_+(\Omega)$ *into* $L^2_+(\Omega)$.

4.5.4   Applications of Krasnoselskii-Type Fixed Point Theorems

By using Corollary 3.31, Corollary 4.21, Theorem 4.20, and the definition of the upper scalar derivatives it can be shown as follows.

THEOREM 4.22 *If conditions 1–6 of Corollary 4.21 and condition*

7. $\exists \varepsilon, \delta > 0$ *such that*

$$\frac{\mathcal{I}_3(\mathcal{L})(s,t,u) - \mathcal{I}_3(\mathcal{L})(s,t,0)}{u} \leq 1 - \delta,$$

for almost all $s,t \in \Omega$ *and for all* $u \in [-\varepsilon,\varepsilon]^\Omega$

are satisfied, then the integral equation

$$u(s) = \int_\Omega \mathcal{L}(s, t, u(t))dt$$

has a solution $u \in L^2_+(\Omega)$.

Proof. Consider the integral operator $\mathcal{A}$ defined by the relation

$$\mathcal{A}(u)(s) = \int_\Omega \mathcal{L}(s, t, u(t))dt.$$

By Corollary 4.21, $\mathcal{A}$ is a completely continuous operator from $L^2_+(\Omega)$ into $L^2_+(\Omega)$. It is easy to see that

$$\mathcal{I}(\mathcal{A})(u)(s) = \int_\Omega \mathcal{I}_3(\mathcal{L})(s, t, u(t))dt. \tag{4.4}$$

By (4.4)

$$\frac{\langle \mathcal{I}(\mathcal{A})(u) - \mathcal{I}(\mathcal{A})(0), u\rangle}{\|u\|^2}$$

$$= \frac{\int_\Omega \int_\Omega (\mathcal{I}_3(\mathcal{L})(s, t, u(t)) - \mathcal{I}_3(\mathcal{L})(s, t, 0))u(s)dsdt}{\int_\Omega u^2(s)ds}$$

$$= \frac{\int_\Omega \int_\Omega \frac{(\mathcal{I}_3(\mathcal{L})(s, t, u(t)) - \mathcal{I}_3(\mathcal{L})(s, t, 0))}{u(t)}u(s)u(t)dsdt}{\int_\Omega u^2(s)ds}.$$

By the Cauchy inequality

$$\int_\Omega \int_\Omega u(s)u(t)dsdt = \left(\int_\Omega u(s)ds\right)^2 \le \int_\Omega u^2(s)ds. \tag{4.5}$$

By using (4.5) and the definition of the upper scalar derivative, we have $\overline{\mathcal{I}(\mathcal{A})}^{\#}(0) < 1$, if 6. holds. Hence, Theorem 4.22 is a consequence of Corollary 3.31 and Theorem 4.20. $\square$

COROLLARY 4.23 *If conditions 1–6 of Corollary 4.21 with $\mathcal{K}(s, t)\mathcal{F}(t, u)$ in place of $\mathcal{L}(s, t, u)$ and condition*

7. $\exists \varepsilon, \delta > 0$ *such that*

$$\mathcal{K}(s, t)\frac{\mathcal{I}_2(\mathcal{F})(t, u) - \mathcal{I}_2(\mathcal{F})(t, 0)}{u} \le 1 - \delta,$$

for almost all $s, t \in \Omega$ *and all* $u \in [-\varepsilon, \varepsilon]^{\Omega}$

are satisfied, then the integral equation

$$u(s) = \int_{\Omega} \mathcal{K}(s, t) \mathcal{F}(t, u(t)) dt$$

has a solution $u \in L_+^2(\Omega)$.

By using Corollary 3.33 it can be proved similarly to Theorem 4.22 and Corollary 4.23 as follows.

THEOREM 4.24 *If conditions 1–6 of Corollary 4.21 with*

$$\frac{1}{\operatorname{mes}(\Omega)} u - \mathcal{L}(s, t, u)$$

in place of $\mathcal{L}(s, t, u)$ *and condition*

7. $\exists \varepsilon, \delta > 0$ *such that*

$$\frac{\mathcal{I}_3(\mathcal{L})(s, t, u) - \mathcal{I}_3(\mathcal{L})(s, t, 0)}{u} \geq \delta,$$

for almost all $s, t \in \Omega$ *and all* $u \in [-\varepsilon, \varepsilon]^{\Omega}$

are satisfied, then the integral equation

$$v(s) = \int_{\Omega} \mathcal{L}(s, t, u(t)) dt$$

has a solution $u \in L_+^2(\Omega)$ *for every* $v \in L_+^2(\Omega)$.

COROLLARY 4.25 *If conditions 1–6 of Corollary 4.21 with*

$$\frac{1}{\operatorname{mes}(\Omega)} u - \mathcal{K}(s, t) \mathcal{F}(t, u)$$

in place of $\mathcal{L}(s, t, u)$ *and condition*

7. $\exists \varepsilon, \delta > 0$ *such that*

$$\mathcal{K}(s, t) \frac{\mathcal{I}_2(\mathcal{F})(t, u) - \mathcal{I}_2(\mathcal{F})(t, 0)}{u} \geq \delta,$$

for almost all $s, t \in \Omega$ *and all* $u \in [-\varepsilon, \varepsilon]^{\Omega}$

are satisfied, then the integral equation

$$v(s) = \int_\Omega \mathcal{K}(s,t)\mathcal{F}(t,u(t))dt$$

has a solution $u \in L^2_+(\Omega)$ *for every* $v \in L^2_+(\Omega)$.

4.5.5 Applications of Altman-Type Fixed Point Theorems

We remark that particularly every completely continuous mapping is α-condensing and the scalar product of a Hilbert space is a semi-inner product compatible with the norm generated by the scalar product. The reader should bear this in mind when reference is made to the results of Section 4.5.

By using Theorem 4.17, Theorem 4.20, and the definition of the upper scalar derivative it can be shown as follows.

THEOREM 4.26 *If conditions 1–5 of Theorem 4.20 and condition*

6. $\exists \varepsilon, \delta > 0$ *such that*

$$\frac{\mathcal{I}_3(\mathcal{L})(s,t,u) - \mathcal{I}_3(\mathcal{L})(s,t,0)}{u} \leq 1 - \delta,$$

for almost all $s, t \in \Omega$ *and for all* $u \in [-\varepsilon, \varepsilon]^\Omega$

are satisfied, then the integral equation

$$u(s) = \int_\Omega \mathcal{L}(s,t,u(t))dt$$

has a solution $u \in L^2(\Omega)$.

Proof. Consider the integral operator $\mathcal{A}$ defined by the relation

$$\mathcal{A}(u)(s) = \int_\Omega \mathcal{L}(s,t,u(t))dt.$$

By Theorem 4.20, $\mathcal{A}$ is a completely continuous operator from $L^2(\Omega)$ into $L^2(\Omega)$. It is easy to see that

$$\mathcal{I}(\mathcal{A})(u)(s) = \int_\Omega \mathcal{I}_3(\mathcal{L})(s,t,u(t))dt. \tag{4.6}$$

By (4.6)

$$\frac{\langle \mathcal{I}(\mathcal{A})(u) - \mathcal{I}(\mathcal{A})(0), u \rangle}{\|u\|^2}$$

$$= \frac{\displaystyle\int_\Omega \int_\Omega (\mathcal{I}_3(\mathcal{L})(s,t,u(t)) - \mathcal{I}_3(\mathcal{L})(s,t,0))u(s)dsdt}{\displaystyle\int_\Omega u^2(s)ds}$$

$$= \frac{\displaystyle\int_\Omega \int_\Omega \frac{(\mathcal{I}_3(\mathcal{L})(s,t,u(t)) - \mathcal{I}_3(\mathcal{L})(s,t,0))}{u(t)}u(s)u(t)dsdt}{\displaystyle\int_\Omega u^2(s)ds} .$$

By the Cauchy inequality

$$\int_\Omega \int_\Omega u(s)u(t)dsdt = \left(\int_\Omega u(s)ds \right)^2 \leq \int_\Omega u^2(s)ds. \qquad (4.7)$$

By using (4.7) and the definition of the upper scalar derivative, we have $\overline{\mathcal{I}(\mathcal{A})}^{\#}(0) < 1$, if 6. holds. Hence, Theorem 4.26 is a consequence of Theorems 4.17 and 4.20. □

COROLLARY 4.27 *If conditions 1–5 of Theorem 4.20 with $\mathcal{K}(s,t)\mathcal{F}(t,u)$ in place of $\mathcal{L}(s,t,u)$ and condition*

6. $\exists \varepsilon, \delta > 0$ *such that*

$$\mathcal{K}(s,t)\frac{\mathcal{I}_2(\mathcal{F})(t,u) - \mathcal{I}_2(\mathcal{F})(t,0)}{u} \leq 1 - \delta,$$

for almost all $s,t \in \Omega$ and all $u \in [-\varepsilon, \varepsilon]^\Omega$

are satisfied, then the integral equation

$$u(s) = \int_\Omega \mathcal{K}(s,t)\mathcal{F}(t,u(t))dt$$

has a solution $u \in L^2(\Omega)$.

By using Theorem 4.18 it can be proved similarly to Theorem 4.26 and Corollary 4.27 as follows.

THEOREM 4.28 *If conditions 1–5 of Theorem 4.20 with*

$$\frac{1}{\text{mes}(\Omega)} u - \mathcal{L}(s, t, u)$$

in place of $\mathcal{L}(s, t, u)$ *and condition*

6. $\exists \varepsilon, \delta > 0$ *such that*

$$\frac{\mathcal{I}_3(\mathcal{L})(s, t, u) - \mathcal{I}_3(\mathcal{L})(s, t, 0)}{u} \geq \delta,$$

for almost all $s, t \in \Omega$ *and all* $u \in [-\varepsilon, \varepsilon]^\Omega$

are satisfied, then the integral equation

$$v(s) = \int_\Omega \mathcal{L}(s, t, u(t)) dt$$

has a solution $u \in L^2(\Omega)$ *for every* $v \in L^2(\Omega)$.

COROLLARY 4.29 *If conditions 1–5 of Theorem 4.20 with*

$$\frac{1}{\text{mes}(\Omega)} u - \mathcal{K}(s, t) \mathcal{F}(t, u)$$

in place of $\mathcal{L}(s, t, u)$ *and condition*

6. $\exists \varepsilon, \delta > 0$ *such that*

$$\mathcal{K}(s, t) \frac{\mathcal{I}_2(\mathcal{F})(t, u) - \mathcal{I}_2(\mathcal{F})(t, 0)}{u} \geq \delta,$$

for almost all $s, t \in \Omega$ *and all* $u \in [-\varepsilon, \varepsilon]^\Omega$

are satisfied, then the integral equation

$$v(s) = \int_\Omega \mathcal{K}(s, t) \mathcal{F}(t, u(t)) dt$$

has a solution $u \in L^2(\Omega)$ *for every* $v \in L^2(\Omega)$.

REMARK 4.3 *We could have considered the closed convex cone* $L_+^2(\Omega)$ *instead of the whole space* $L^2(\Omega)$. *But in this case we would have obtained results already presented in the previous part, because* $L_+^2(\Omega)$ *is a generating closed convex pointed cone of* $L^2(\Omega)$. *The results corresponding to Theorem 4.18 would have been even more particular, because of the invariance condition* $(I - f)(K) \subset K$.

Chapter 5

Monotone Vector Fields on Riemannian Manifolds and Scalar Derivatives

In this chapter we generalize the notion of Kachurovskii–Minty–Browder monotonicity (see [Browder, 1964, Kachurovskii, 1960, 1968; Minty, 1962, 1963]) to Riemannian manifolds. The results of this chapter are based on the papers [Nemeth, 1999a,b,c, 2001]. For global examples of monotone vector fields we often consider Hadamard manifolds (complete, simply connected Riemannian manifolds of nonpositive sectional curvature). However, these examples can be extended locally for positive curvature manifolds, by using the comparison theorems. To avoid technical difficulties we consider just the Hadamard manifolds, where these vector fields can be defined globally. First, we fix the notions and results used throughout this chapter. If M is a manifold and $a \in M$, we denote the tangent space of M in a by $T_a M$. If $A \subset M$, by a vector field on A we mean a map

$$A \ni a \longmapsto V_a \in T_a M.$$

Let M be endowed by a Riemannian metric $g(\cdot, \cdot)$, with corresponding norm denoted $\|\cdot\|$. The same notation is used for the norm of a Euclidean space. We suppose that the notions and the fundamental properties of the gradient of a vector field on a manifold (see [Spivak, 1965]), parallel transport, geodesic, geodesic segment, geodesic distance function, and exponential map on a Riemannian manifold (see [do Carmo, 1992]) are known. We also suppose that the reader is familiar with the Hopf–Rinow theorem for Riemannian manifolds (see [do Carmo, 1992]), the Hadamard theorem for Hadamard manifolds (see [O'Neil, 1983]), and the fixed point theorem of Brouwer (see [Zeidler, 1986]). We denote the geodesic distance function of a Riemannian manifold by d, the exponential map of a Riemannian manifold by "exp" and the inverse map of the exponential map of a Hadamard manifold by $\exp^{-1}$ (which is well defined by the Hadamard theorem).

Because in our results it is important to make a difference between the tangent vector of a geodesic γ with respect to the arclength and the tangent vector of a geodesic γ with respect to an arbitrary parameter we denote the former $\dot{\gamma}$ and the latter γ', respectively.

5.1   Geodesic Monotone Vector Fields

Although many results of convex analysis and optimization theory were generalized to Riemannian manifolds, there are few attempts to generalize the Kachurovskii–Minty–Browder monotonicity notion (see [Browder, 1964; Kachurovskii, 1960, 1968; Minty, 1962, 1963]).

C. Udrişte proves [Udrişte, 1976] that a function f defined on an open and (geodesic) convex subset [Rapcsak, 1997] of a Riemannian manifold is (geodesic) convex [Rapcsak, 1997], if and only if its gradient is geodesic monotone. This result is expressed, in terms of the differential of f, and is stated in [Udrişte, 1977] too. [T. Rapcsak, 1997] reformulates the result of Udrişte, giving explicitly the inequality which express the geodesic monotonicity of the gradient of a convex function. However, as far as we know the nongradient type geodesic monotone vector fields have not been studied yet. Connected to the notion of geodesic monotonicity we introduce the strictly geodesic monotone, virtually geodesic monotone, and trivially geodesic monotone vector fields.

We prove that a vector field X on a Riemannian manifold is geodesic monotone (strictly geodesic monotone) if and only if the first variation of the length of every geodesic arc γ with infinitesimal variation the restriction of X to γ is nonnegative (positive).

Analysing the existence of geodesic monotone vector fields, we prove that there is no strictly geodesic monotone vector field on a Riemannian manifold which contains a closed geodesic. Similarly, if every geodesic of a Riemannian manifold is closed then there is no virtually monotone vector field on the manifold. As a consequence, because every compact and complete Riemannian manifold contains a closed geodesic [Klingenberg, 1978], there are no strictly monotone vector fields on compact and complete Riemannian manifolds. For the existence of strictly geodesic monotone vector fields, we consider Hadamard manifolds (simply connected, complete Riemannian manifolds with nonpositive sectional curvature), proving that for f strictly monotone, the f-position vector fields of such manifolds are strictly geodesic monotone, where the f-position vector fields generalize the notion of position vector fields.

We can generalize the notion of the scalar derivative (see Sections 5.1.1 and 5.1.2), introduced by us for characterizing such vector fields. The results are generalizations of some results of Sections 5.1.1 and 5.1.2.

The case of constant sectional curvature Riemannian manifolds presents some interesting peculiarities, therefore we consider it separately. For illustrating the ideas it is enough to consider just the cases of $\mathbb{S}^n$ and $\mathbb{H}^n$, where $\mathbb{S}^n$ and $\mathbb{H}^n$ are the n-dimensional unit sphere and hyperbolical space respectively. Of course similar results hold for every constant curvature manifold.

5.1.1 Geodesic Monotone Vector Fields and Convex Functionals

DEFINITION 5.1 *Let M be a Riemannian manifold.*

1. *A subset K of M is called* (geodesic) convex [Rapcsak, 1997] *if for any two points of M there is a geodesic arc contained in K joining these points.*

2. *Let K be a convex subset of M. A function $f : K \to \mathbb{R}$ is called* (geodesic) convex [Rapcsak, 1997] *(strictly (geodesic) convex [Rapcsak, 1997]), if $f \circ \gamma : [0, l] \to \mathbb{R}$ is convex (strictly convex) for every unit speed geodesic arc $\gamma : [0, l] \to M$ contained in K.*

If N is an arbitrary manifold, we denote by $\mathrm{Sec}(TN)$ the family of sections of the tangent bundle TN of N. By using this notation we have the following definition.

DEFINITION 5.2 *Let (M, g) be a Riemannian manifold, $K \subset M$ a geodesic convex open set, and $X \in \mathrm{Sec}(TK)$ a vector field on K.*

1. *X is called* geodesic monotone *if for every $x, y \in K$ and every unit speed geodesic arc $\gamma : [0, l] \to M$ joining x and y ($\gamma(0) = x, \gamma(l) = y$) and contained in K we have that*

$$g(X_x, \dot{\gamma}(0)) \leq g(X_y, \dot{\gamma}(l)),$$

where $\dot{\gamma}$ denotes the tangent vector of γ with respect to the arclength.

2. *X is called* strictly geodesic monotone *if for every distinct x and y we have*

$$g(X_x, \dot{\gamma}(0)) < g(X_y, \dot{\gamma}(l)).$$

3. *X is called* virtually geodesic monotone *if it is geodesic monotone and there are x and y such that*

$$g(X_x, \dot{\gamma}(0)) < g(X_y, \dot{\gamma}(l)).$$

4. *X is called* trivially geodesic monotone *if for every x, y we have that*

$$g(X_x, \dot{\gamma}(0)) = g(X_y, \dot{\gamma}(l)).$$

Because the length of the tangent vector of an arbitrary parametrized geodesic is constant, the relations of Definition 5.2 can be given for any parametrization of γ. It is also easy to see that X is geodesic monotone (strictly geodesic monotone), if and only if for every arbitrarily parametrized geodesic γ the function

$$v : \tau \mapsto g(X_{\gamma(\tau)}, \gamma'(\tau))$$

is monotone (strictly monotone), where $\gamma'(\tau)$ is the tangent vector of γ, with respect to its parameter τ. Similarly X is trivially monotone, if and only if v is constant.

The following example makes the connection between geodesic monotone vector fields and monotone operators of an Euclidean space, showing that with few modifications the former are generalization of the latter.

EXAMPLE 5.3 *Let E be an Euclidean space, $G \subset E$ an open and convex set, and $h : G \to E$ a monotone operator. Then the vector field $X \in \mathrm{Sec}(TG)$; $x \mapsto h(x)_x$, where $h(x)_x$ is the tangent vector in 0 of the curve $t \mapsto x + th(x)$, is geodesic monotone.*

The following theorem is a modified version of Udriṣte's result [Udriṣte, 1976, 1977].

THEOREM 5.4 *Let M be a Riemannian manifold and K an open and convex subset of M. A function $f : K \to \mathbb{R}$ is convex (strictly convex), if and only if its gradient $\mathrm{grad} f$ is geodesic monotone (strictly geodesic monotone) [Udriṣte, 1976].*

Udriṣte gives the inequality which expresses the geodesic monotonicity in terms of df, the differential of f.

Rapcsák [1997] states the inequality in explicit form, by using the gradient of f. However, neither of them speaks (in general) about monotone vector fields.

5.1.2 Geodesic Monotone Vector Fields and the First Variation of the Length of a Geodesic

In this section we make a connection between the geodesic monotone vector fields and the first variation of the lengths of geodesics. First of all we recall some preliminary definitions and results.

DEFINITION 5.5 *Let (M, g) be a Riemannian manifold and $\xi : [\lambda, \nu] \to M$ a piecewise smooth continuous curve with breaking points*

$$\lambda < t_1 < \cdots < t_{n-1} < \nu.$$

Let $t_0 = \lambda$ and $t_n = \nu$. A piecewise smooth deformation of the curve ξ is a continuous map

$$\delta : [\lambda, \nu] \times [-\varepsilon, \varepsilon] \to M; \quad \varepsilon > 0,$$

which is smooth on the rectangles $[t_{i-1}, t_i] \times [-\varepsilon, \varepsilon]$; $i = \overline{1, n}$ *and* $\delta(t, 0) = \xi(t)$; $\forall t \in [\lambda, \nu]$. *Let* s *be the parameter which varies in* $[-\varepsilon, \varepsilon]$. *The vector field* V *along* ξ *defined by*

$$V(t) = \frac{d\delta}{ds}|_{s=0}$$

the infinitesimal deformation *induced by* δ. *For a fixed* $s \in [-\varepsilon, \varepsilon]$, *the curve* $\xi_s(t) = \delta(t, s)$ *is called the* deformed curve; *its arclength is a smooth function*

$$L(s) = \int_\lambda^\nu \|\xi_s'(t)\| dt.$$

We call the first variation *of the arclength of* ξ *corresponding to the deformation* δ *the number*

$$L'(0) = \frac{dL}{ds}|_{s=0}.$$

DEFINITION 5.6 *Let* $\xi : [\lambda, \nu] \to M$ *be a piecewise smooth curve of a Riemannian manifold* (M, g) *and* $\lambda < t_1 < \ldots < t_{n-1} < \nu$ *its breaking points. Then the vectors*

$$\Theta\xi'(t_i) = \xi'(t_i^+) - \xi'(t_i^-) \in T_{\xi(t_i)}M; \ i = \overline{1, n-1}$$

are called the breakings *of the tangent vector.*

Then we have the following proposition.

PROPOSITION 5.1 *Let* $\xi : [\lambda, \nu] \to M$ *be a piecewise smooth curve of a Riemannian manifold, parametrized by arclength* $\lambda < t_1 < \cdots < t_{n-1} < \nu$ *its breaking points,* δ *a piecewise smooth deformation of* ξ, *and* $V : [\lambda, \nu] \to TM$ *the infinitesimal deformation induced by* δ. *Then the first variation of the length of* ξ *is*

$$L'(0) = -\int_\lambda^\nu g(\ddot{\xi}(t), V(t)) dt - \sum_{i=1}^{n-1} g(\Theta\dot{\xi}(t_i), V(t_i)) + g(\dot{\xi}, V)|_\lambda^\nu. \quad (5.1)$$

The next theorem is the main result of this section, connecting the notion of geodesic monotonicity with the first variations of the lengths of geodesics:

THEOREM 5.7 *Let* (M, g) *be a Riemannian manifold,* K *an open and geodesic convex subset of* M, *and* X *a smooth vector field on* K. *Then,* X *is geodesic monotone (strictly geodesic monotone), if and only if for all unit speed geodesic arcs* $\gamma : [0, l] \to M$ *contained in* K *and all deformation* δ *inducing the infinitesimal variation* V *equal to the restriction of* X *to* γ, *the first variation of the length of* γ *is nonnegative (positive).*

Proof. Let x, y be two arbitrary, distinct points of M and $\gamma : [0, l] \to M$ a unit speed geodesic arc contained in K and joining x, y, such that $\gamma(0) = x$ and $\gamma(l) = y$. Let δ be a deformation of γ inducing the infinitesimal variation V equal to the restriction of X to γ. Because γ is a geodesic, the first and second terms in the variation formula (5.1) vanish. Hence we have for the first variation of the length of γ:

$$L'(0) = g(\dot{\gamma}, V)|_0^l.$$

V is the restriction of X to γ, and thus we have that

$$L'(0) = g(\dot{\gamma}(l), X_y) - g(\dot{\gamma}(0), X_x). \tag{5.2}$$

By using the definition of geodesic monotonicity (strict geodesic monotonicity), (5.2) proves the theorem.   □

5.1.3   Closed Geodesics and Geodesic Monotone Vector Fields

In this section we analyse the connection between the existence of closed geodesics and geodesic monotone vector fields. We start with the following theorem.

THEOREM 5.8 *Let (M, g) be a Riemannian manifold which contains a closed geodesic γ. Then there are no strictly geodesic monotone vector fields on M.*

Proof. Suppose that there is a strictly geodesic monotone vector field X on M. Let τ be the parameter of γ and $x = \gamma(\tau_1)$, $y = \gamma(\tau_2)$ two arbitrary points of the curve. Then we have

$$g(X_x, \gamma'(\tau_1)) < g(X_y, \gamma'(\tau_2)) \tag{5.3}$$

and

$$g(X_y, \gamma'(\tau_2)) < g(X_x, \gamma'(\tau_1)), \tag{5.4}$$

because $g(X_{\gamma'(\tau)}, \gamma'(\tau))$ is strictly monotone and γ is closed, respectively. But inequalities (5.3) and (5.4) imply a contradiction. Hence there are no strictly geodesic monotone vector fields on M.   □

By using the same argument as in the proof of Theorem 5.8 we obtain the following result.

THEOREM 5.9 *If every geodesic of a Riemannian manifold is closed then there are no virtually monotone vector fields on the manifold.*

Because every compact and complete Riemannian manifold contains a closed geodesic [Klingenberg, 1978], we have the following theorem.

THEOREM 5.10 *There are no strictly geodesic monotone vector fields on compact and complete Riemannian manifolds.*

5.1.4 The Geodesic Monotonicity of Position Vector Fields

In this section we define the notion of f-position vector fields of a Riemannian manifold M and analyse their geodesic monotonicity under certain topological and metrical restrictions imposed on M.

Recall that a complete, simply connected Riemannian manifold, of nonpositive sectional curvature is called an *Hadamard manifold*.

By the Gauss lemma [O'Neil, 1983, p. 127] it can be seen easily that the following definition generalizes the notion of position vector fields [O'Neil, 1983, p. 178] on Hadamard manifolds. (The definition is given just for local position vector fields. However by the theorem of Hadamard [O'Neil, 1983, p. 278] these vector fields can be defined globally on Hadamard manifolds.)

DEFINITION 5.11 *Let M be an Hadamard manifold and $o \in M$. If $x \neq o$, denote by l_x the distance of x from o, and by γ_x the unit speed radial geodesic joining o with x; $\gamma_x(0) = o$. Let $f : [0, \infty[\to [0, \infty[$. Then the vector field $P^{f,o}$ defined by*

$$P^{f,o}_x = \begin{cases} f(l_x)\dot{\gamma}_x(l_x) & \text{if } x \neq o, \\ 0 & \text{if } x = o. \end{cases}$$

is called the f-position vector field of M at o.

In particular, the position vector field at o of an Hadamard manifold is equal to the id-position vector field of M at o, where id denotes the identity map of $[0, \infty[$. The local f-position vector field at a point of an arbitrary Riemannian manifold (in a normal neighbourhood) can be defined similarly.

In the following definition indices $i = \overline{1,3}$ are considered modulo 3:

DEFINITION 5.12 *A geodesic triangle T in a Riemannian manifold M is a set formed by three segments of minimizing unit speed geodesics (called* sides *of the triangle)*

$$\gamma_i : [0, l_i] \to M; \ i = \overline{1,3},$$

in such a way that $\gamma_i(l_i) = \gamma_{i+1}(0)$; $i = \overline{1,3}$. The endpoints of the geodesic segments are called vertices *of T. The angle*

$$\angle(-\dot{\gamma}_i(l_i), \dot{\gamma}_{i+1}(0)); \ i = \overline{1,3}$$

is called the (interior) angle of the corresponding vertex.

Then we have the following lemma, proved in [do Carmo, 1992, p. 259].

LEMMA 5.13 *Let M be an Hadamard manifold of sectional curvature κ. Let a, b, and c be three points of M. Then (by the Hadamard theorem [O'Neil, 1983, p. 278]) such points determine a unique geodesic triangle T in M with vertices* *a, b, c. Let α, β, and γ be the angles of the vertices a, b, c, respectively, and let A, B, C be the lengths of the sides opposite the vertices a, b, c, respectively. Then*

1. $A^2 + B^2 - 2AB\cos\gamma \leq C^2$ *(< C^2, if $\kappa < 0$).*

2. $\alpha + \beta + \gamma \leq \pi$ *(< π, if $\kappa < 0$).*

COROLLARY 5.14 *Let M be an Hadamard manifold and a, b, and c three points of M. Then (by the Hadamard theorem [O'Neil, 1983, p. 278]) such points determine a unique geodesic triangle T in M with vertices a, b, c. Let α, β, and γ be the angles of the vertices a, b, c, respectively, and let A, B, C be the lengths of the sides opposite the vertices a, b, c, respectively. If $f : [0, \infty[\to [0, \infty[$ is a strictly increasing function, then we have*

$$f(C)\cos\beta + f(B)\cos\gamma > 0.$$

Proof. By 2. of Lemma 5.13 we have that $\beta + \gamma < \pi$. Hence either $\beta < \pi/2$, or $\gamma < \pi/2$. Without losing generality we can suppose that $\beta \leq \gamma$. We consider two cases:

1. $0 < \beta \leq \gamma < \pi/2$. In this case the inequality of the lemma is trivial.

2. $\gamma \geq \pi/2$ and $\beta < \pi/2$. In this case we have $\cos\gamma < 0$. Hence by 1. of Lemma 5.13 we have that

$$B^2 < A^2 + B^2 - 2AB\cos\gamma \leq C^2.$$

Thus we get $0 < B < C$, and consequently $0 \leq f(0) < f(B) < f(C)$, because f is strictly increasing. By using 2 of Lemma 5.13 and $\cos\beta > 0$ we obtain that

$$f(C)\cos\beta + f(B)\cos\gamma > f(B)(\cos\beta + \cos\gamma) > 0.$$

$\square$

THEOREM 5.15 *Let M be an Hadamard manifold and $f : [0, \infty[\to [0, \infty[$ a strictly increasing function. Then for all $o \in M$ the f-position vector field at o is strictly geodesic monotone.*

Proof. Let $x \neq y \in M$.

1. If x, y belongs to the same geodesical ray (with respect to o), then we trivially have that

$$g(P_x^{f,o}, \dot\gamma(0)) < g(P_x^{f,o}, \dot\gamma(l)),$$

where $\gamma : [0, l] \to M$ is the unit speed geodesic arc joining x and y, such that $\gamma(0) = x$ and $\gamma(l) = y$.

2. If x, y do not belong to the same geodesical ray the above inequality can be easily deduced by using Corollary 5.14.

□

In particular the position vector field at o is strictly geodesic monotone. However, the monotonicity of this vector field is an easy consequence of Theorem 5.7, Gauss' lemma [O'Neil, 1983, p. 127], Corollary 3. [O'Neil, 1983, p. 128], and the convexity of the distance function from a point in an Hadamard manifold.

We remark that, if f is bounded above, then P^f is a smooth, strictly geodesic monotone and bounded vector field. (For example, we can take the function $f = \arctan |_{[0,\infty[}$.)

LEMMA 5.16 *Let* $\phi : [0, \infty[\to [0, \infty[$ *be a differentiable function with* $\phi(0) = 0$. *Then there exists a differentiable function* $\psi : [0, \infty[\to [0, \infty[$, *with* $\psi(0) = (d\phi/dt)(0)$, $\phi(t) = t\psi(t)$, $t \in [0, 1]$.

Proof. It suffices to define, for fixed t

$$\psi(t) = \int_0^1 \frac{d\phi(ts)}{d(ts)} ds$$

and, after changing the variables, observe that

$$t\psi(t) = \int_0^t \frac{d\phi(ts)}{d(ts)} d(ts) = \phi(t).$$

□

By applying Lemma 5.16 to the function f of Theorem 5.15, we have that there is a smooth function $h : [0, \infty[\to [0, \infty[$, with $h(0) = (df/dt)(0)$ such that $f(t) = th(t)$. Hence, by using Definition 5.11, we have the following proposition.

PROPOSITION 5.2 *Let* M *be an Hadamard manifold and* $o \in M$. *Let* $f : [0, \infty[\to [0, \infty[$ *be a smooth function with* $f(0) = 0$ *and* $h : [0, \infty[\to [0, \infty[$ *the smooth function defined by*

$$\begin{cases} h(t) = \dfrac{f(t)}{t} & \text{if } t \neq 0, \\[2mm] h(0) = \dfrac{df}{dt}(0). \end{cases}$$

Then the f-position vector field $P^{f,o}$ of M at o assigns to each $x \in M$, the vector

$$P_x^{f,o} = h(l_x)P_x^o,$$

where P^o is the position vector field at o and l_x is the distance of x from o.

PROPOSITION 5.3 We use the notations of Proposition 5.2. Let f be strictly monotone and $k : M \to [0, \infty[$ the smooth function defined by $k(x) = h(l_x)$. Then a necessary and sufficient condition for $P^{f,o}$ to be the gradient of a strictly convex function is

$$dk \wedge P^{o^\flat} = 0, \tag{5.5}$$

where for a smooth vector field X on M, $X^\flat$ is the smooth 1-form defined by

$$X^\flat(Y) = g(X, Y),$$

for every smooth vector field Y on M.

If the integrability condition (5.5) is satisfied, then $P^{f,o} = \operatorname{grad} \mu$, where μ is the strictly convex function defined by

$$\mu(x) = \int_o^x kP^{o^\flat}, \tag{5.6}$$

where the integral is taken along any curve joining o with x (e.g., we can take it along γ_x, the radial geodesic joining o with x).

Proof. Suppose that $P^{f,o} = \operatorname{grad} \mu$, where $\mu : M \to \mathbb{R}$. By Theorem 5.4 μ must be strictly convex, because $P^{f,o}$ is strictly geodesic monotone. Then $P^{f,o^\flat}(Y) = g(\operatorname{grad} \mu, Y)$. Hence $P^{f,o^\flat} = d\mu$. Let $\tilde{q} : T_o(M) \to \mathbb{R}$; $\tilde{q}(v) = g(v, v)$, and $q = \tilde{q} \circ \exp_o^{-1}$. Because $2P = \operatorname{grad} q$ (by Corollary 3. of [O'Neil, 1983, p. 128]), a similar argument proves that $2P^{o^\flat} = dq$. Hence $P^{o^\flat}$ is exact. By Proposition 5.2 we have that

$$P^{f,o^\flat} = kP^{o^\flat}. \tag{5.7}$$

Because $P^{f,o^\flat}$ is exact, it is closed. Hence (5.7) implies

$$0 = dP^{f,o^\flat} = dk \wedge P + k dP^{o^\flat}.$$

But $dP^{o^\flat} = 0$, because $P^{o^\flat}$ is exact. Hence we have that

$$dk \wedge P = 0.$$

Inasmuch as M is simply connected its first cohomology group $H^1(M, \mathbb{R}) = 0$. Thus if the integrability condition (5.5) is satisfied (i.e, $P^{f,o^\flat}$ is closed) then $P^{f,o^\flat}$ is exact and

$$P^{f,o^\flat} = d\mu,$$

where

$$\mu(x) = \int_o^x P^{f,o^b} = \int_o^x k P^{o^b}. \qquad (5.8)$$

The first equality of (5.8) was obtained by using formula (6.50) p. 130 of [Nash and Sen, 1983]. Hence we have that $P^f = \text{grad } \mu$, with μ given by (5.6). $\square$

Hence, if the integrability condition $dk \wedge P = 0$ is not satisfied, $P^{f,o}$ cannot be obtained as a gradient of a strictly convex function. This shows that there are strictly geodesic monotone vector fields on an Hadamard manifold which cannot be obtained as gradients of strictly convex functions.

5.1.5 Geodesic Scalar Derivative

In this section we generalize the notion of the scalar derivative, and some of our results concerning the characterization of monotone operators through the scalar derivative, included in Sections 1.1.2 and 1.1.5 (see also [Nemeth, 1992, 1993]).

Let us start by defining the geodesic scalar derivative of a vector field.

DEFINITION 5.17 *Let (M, g) be a Riemannian manifold, $K \subset M$ a convex open set, and $X \in \text{Sec}(TK)$ a vector field on K. Then the* lower geodesic scalar derivative *(upper geodesic scalar derivative) of X is the function*

$$\underline{X}^{\#} : K \to \mathbb{R}; \quad \underline{X}^{\#}(x) = \liminf_{\substack{t \downarrow 0 \\ \gamma \in \Gamma}} \frac{g(X_{\gamma(t)}, \dot{\gamma}(t)) - g(X_x, \dot{\gamma}(0))}{t}$$

$$\left(\overline{X}^{\#} : K \to \mathbb{R}; \quad \overline{X}^{\#}(x) = \limsup_{\substack{t \downarrow 0 \\ \gamma \in \Gamma}} \frac{g(X_{\gamma(t)}, \dot{\gamma}(t)) - g(X_x, \dot{\gamma}(0))}{t} \right),$$

where Γ denotes the family of unit speed geodesic arcs $\gamma : [0, l] \to M$ starting from x (i.e., $\gamma(0) = x$), and contained in K.

If $x_0 \in K$ and $\underline{X}^{\#}(x_0) = \overline{X}^{\#}(x_0) =: X^{\#}(x_0)$, then X is called geodesic scalarly differentiable at x_0 *and $X^{\#}(x_0)$ is called the* geodesic scalar derivative *of X in x_0. If X is geodesic scalarly differentiable at every $x \in K$ then X is called* geodesic scalarly differentiable, *and the function $X^{\#} : x \mapsto X^{\#}(x)$ is called the* geodesic scalar derivative *of X.*

The following theorem is a local characterization of geodesic monotone vector fields.

THEOREM 5.18 *Let (M, g) be a Riemannian manifold, $K \subset M$ a convex open set, and $X \in \text{Sec}(TK)$ a vector field on K. Then we have the following assertions.*

1. $X(-X)$ is geodesic monotone if and only if $\underline{X}^{\#}(x) \geq 0$ $(\overline{X}^{\#}(x) \leq 0)$ for all $x \in K$.

2. X is trivially monotone if and only if X is geodesic scalarly differentiable and $X^{\#}(x) = 0$ for all $x \in K$.

Proof.

1. First we suppose that $\underline{X}^{\#}(x) \geq 0$ for every $x \in K$ and prove that X is geodesic monotone. Let a and b be two arbitrary distinct points of K and $\gamma : [0, l] \to M$ a unit speed geodesic arc joining a and b ($\gamma(0) = a$, $\gamma(l) = b$) contained in K. Let $x = \gamma(t)$ be an arbitrary point of the geodesic arc. Let $\varepsilon > 0$ be an arbitrary but fixed positive number. Because $\underline{X}^{\#}(x) \geq 0$, there is a $\delta(t) > 0$ such that for every $s \in I_t =]t - \delta(t), t + \delta(t)[$ we have that

$$\frac{g(X_{\gamma(s)}, \dot{\gamma}(s)) - g(X_x, \dot{\gamma}(t))}{s - t} > -\frac{\varepsilon}{l}. \tag{5.9}$$

(The geodesic arc can be continued to an open one contained in K and $\delta(t)$ chosen sufficiently small so that $\gamma(s)$ can be defined.) But $\{I_t : t \in [0, l]\}$ is an open covering of the compact set $[0, l]$. Hence

$$[0, l] \subset I_{t_1} \cup I_{t_2} \cup \cdots \cup I_{t_{m-1}}$$

for some positive integer m and some points $t_1 < t_2 < \cdots < t_{m-1}$ of $[0, l]$. This yields $0 =: t_0 \in I_{t_1}$ and $l =: t_m \in I_{t_{m-1}}$. Obviously we can choose the intervals $\{I_i : i = \overline{1, m-1}\}$ so that no interval is contained in another. Let $\xi_i \in I_{t_{i-1}} \cap I_{t_i} \cap]t_{i-1}, t_i[$ for $i = \overline{1, m}$. Then, by using (5.9) we have that

$$g(X_{\gamma(\xi_i)}, \dot{\gamma}(\xi_i)) - g(X_{\gamma(t_{i-1})}, \dot{\gamma}(t_{i-1})) > -\frac{\varepsilon}{l}(\xi_i - t_{i-1}), \tag{5.10}$$

and

$$g(X_{\gamma(t_i)}, \dot{\gamma}(t_i)) - g(X_{\gamma(\xi_i)}, \dot{\gamma}(\xi_i)) > -\frac{\varepsilon}{l}(t_i - \xi_i), \tag{5.11}$$

for $i = \overline{1, m}$. Summing the inequalities (5.10) and (5.11) we obtain

$$g(X_{\gamma(t_i)}, \dot{\gamma}(t_i)) - g(X_{\gamma(t_{i-1})}, \dot{\gamma}(t_{i-1})) > -\frac{\varepsilon}{l}(t_i - t_{i-1}), \tag{5.12}$$

for $i = \overline{1, m}$. Summing the inequalities (5.12) for $i = \overline{1, m}$ we get

$$g(X_{\gamma(l)}, \dot{\gamma}(l)) - g(X_{\gamma(0)}, \dot{\gamma}(0)) > -\varepsilon.$$

Because ε is an arbitrary positive number we have that

$$g(X_{\gamma(l)}, \dot{\gamma}(l)) - g(X_{\gamma(0)}, \dot{\gamma}(0)) \geq 0.$$

Hence for any two distinct points a and b of K and an arbitrary geodesic arc joining a and b contained in K, we have that

$$g(X_b, \dot{\gamma}(l)) \geq g(X_a, \dot{\gamma}(0)).$$

Thus X is geodesic monotone. Now suppose that X is geodesic monotone. Then by the definitions of geodesic monotonicity and the geodesic scalar derivative, we have that $\underline{X}^{\#}(x) \geq 0$ for all $x \in K$. The statement of 1 for $-X$ is obtained by using the identity

$$\underline{(-X)}^{\#} = -\overline{X}^{\#}.$$

$\square$

2. We have that X is trivially geodesic monotone if and only if X and $-X$ is geodesic monotone. Hence X is trivially geodesic monotone, if and only if $0 \leq \underline{X}^{\#}(x) \leq \overline{X}^{\#}(x) \leq 0$, for all $x \in K$. Hence $\underline{X}^{\#}(x) = \overline{X}^{\#}(x) = 0$. Thus X is trivially geodesic monotone if and only if it is geodesic scalarly differentiable and $X^{\#}(x) = 0$ for every $x \in K$.

$\square$

Similarly to Theorem 5.18 we can prove the following theorem.

THEOREM 5.19 *Let (M, g) be a Riemannian manifold, $K \subset M$ a convex open set, and $X \in \mathrm{Sec}(TK)$ a vector field on K. If $\underline{X}^{\#}(x) > 0$ for every $x \in K$; then X is strictly geodesic monotone.*

REMARK 5.1 *Theorems 5.18 and 5.19 hold for uncontinuous vector fields too.*

For the smooth case we have the following corollary of Theorems 5.18 and 5.19:

COROLLARY 5.20 *Let (M, g) be a Riemannian manifold, ∇ the Levi–Civitá connection of M, $K \subset M$ a convex set, and $X \in \mathrm{Sec}(TK)$ a smooth vector field on K. For $x \in K$ define $\phi(x) : T_x(M) \to T_x(M)$ by $\phi(x)(v) = \nabla_v X$ for $v \in T_x(M)$, where $T_x(M)$ is the tangent space of M in x. Then we have the following assertions.*

1. $\underline{X}^{\#}(x) = \inf_{\substack{v \in T_x(M) \\ \|v\|=1}} g(\phi(x)(v), v).$

2. *X is geodesic monotone if and only if $\phi(x)$ is positive semi-definite for every x in K.*

3. *X is trivially monotone if and only if $\phi(x)$ is skew-symmetric for every $x \in K$.*

4. *If $\phi(x)$ is positive definite for every $x \in K$, then X is strictly geodesic monotone.*

Proof. Let x be an arbitrary point of K. Then we have

$$\underline{X}^{\#}(x) = \inf_{\gamma \in \Gamma} \liminf_{t \downarrow 0} \frac{g(X_{\gamma(t)}, \dot{\gamma}(t)) - g(X_x, \dot{\gamma}(0))}{t}. \tag{5.13}$$

Equation (5.13) can be rewritten as

$$\underline{X}^{\#}(x) = \inf_{\gamma \in \Gamma} \frac{d}{dt}|_{t=+0}\, g(X_{\gamma(t)}, \dot{\gamma}(t)). \tag{5.14}$$

By using that ∇ is the Levi–Civitá connection of M and γ is geodesic, (5.14) becomes

$$\underline{X}^{\#}(x) = \inf_{\gamma \in \Gamma} g(\nabla_{\dot{\gamma}(0)} X|_x, \dot{\gamma}(0)). \tag{5.15}$$

Because for every $v \in T_x(M)$ there is $\gamma \in \Gamma$ such that $\dot{\gamma}(0) = v$, from (5.15) we obtain that

$$\underline{X}^{\#}(x) = \inf_{\substack{v \in T_x(M) \\ \|v\|=1}} g(\phi(x)(v), v), \tag{5.16}$$

where $\| \cdot \|$ is the norm generated by the metrical tensor of M. By using 1, from (5.16) and Theorems 5.18 and 5.19 we obtain assertions 2, 3 and 4 of the corollary. $\square$

By using Corollary 5.20 we obtain the following theorem.

THEOREM 5.21 *Let (M, g) be a Riemannian manifold, $K \subset M$ a convex open set, and $X \in \mathrm{Sec}(\mathcal{T}K)$ a smooth vector field on K. Then the following assertions are equivalent.*

1. *X is trivially geodesic monotone.*

2. *X is a Killing vector field.*

Proof. By 3 of Corollary 5.20 we have that X is trivially geodesic monotone if and only if A_Y is skew-symmetric, where A_Y is the endomorphism of smooth vector fields defined by $A_Y(X) = \nabla_Y X$. But this is exactly a necessary and sufficient condition for X to be a Killing vector field. $\square$

5.1.6   Geodesic Monotone Vector Fields on $\mathbb{S}^n$

Denote by $\langle \cdot, \cdot \rangle$ the canonical scalar product of $\mathbb{R}^{n+1}$. Consider $\mathbb{S}^n$, the standard unit sphere, to be a Riemannian manifold, with the metric induced from $\mathbb{R}^{n+1}$ (we denote its metric by $\langle \cdot, \cdot \rangle$ too). We start this section by remarking that there are no virtually monotone vector fields on the whole sphere. This is an obvious consequence of Theorem 5.9, because every geodesic of the sphere is closed. Hence we can seek virtually monotone vector fields just on convex proper subsets of the sphere. Denote by $\underset{y \to x}{\overset{g}{}}$ the convergence of y to x along geodesics. Then we have the following theorem.

THEOREM 5.22 *Let K be an open and convex subset of $\mathbb{S}^n$, X a vector field on K, and x an arbitrary point of K. Then we have the following formula for the lower geodesic scalar derivative of X in x,*

$$\underline{X}^{\#}(x) = \liminf_{\overset{g}{y \to x}} \frac{\langle X_y - X_x, y - x \rangle}{\|y - x\|^2}.$$

Proof. Every unit speed geodesic of the n-dimensional unit sphere $\mathbb{S}^n$ starting from $x \in \mathbb{S}^n$ ($\gamma(0) = x$) has equation:

$$\gamma(t) = (\cos t)x + (\sin t)e, \tag{5.17}$$

where $e = \dot{\gamma}(0) \in T_x(\mathbb{S}^n)$ is the tangent unit vector of γ in the starting point. We want to find the equation of the unit speed geodesic, denoted with the same letter γ, starting from $x \in \mathbb{S}^n$ and passing through $y \in \mathbb{S}^n$. Obviously this is equivalent to finding the tangent unit vector e of γ in the starting point. This can be calculated in the following way.

First we find the parameter l corresponding to y and after that we can compute the unit tangent vector of γ: $e = \dot{\gamma}(0)$ by using (5.17). Having in mind that x and y are unit vectors of $T_0(\mathbb{R}^{n+1})$, the tangent space of $\mathbb{R}^{n+1}$ in the origin, we have that

$$\cos l = \langle x, y \rangle. \tag{5.18}$$

Thus we get

$$e = \pm \frac{y - \langle x, y \rangle x}{\sqrt{1 - \langle x, y \rangle^2}}. \tag{5.19}$$

From (5.17) through (5.19) we have that

$$\dot{\gamma}(l) = \mp\sqrt{1 - \langle x, y \rangle^2}\, x \pm \langle x, y \rangle \frac{y - \langle x, y \rangle x}{\sqrt{1 - \langle x, y \rangle^2}}. \tag{5.20}$$

If Γ has the same meaning as in the previous chapter, then we have

$$\underline{X}^{\#}(x) = \liminf_{\substack{l \downarrow 0 \\ \gamma \in \Gamma}} \frac{\langle X_{\gamma(l)}, \dot{\gamma}(l) \rangle - \langle X_{\gamma(0)}, \dot{\gamma}(0) \rangle}{l}. \tag{5.21}$$

Because $\liminf_{l \to 0}(\sin l)/l = 1$, (5.21) becomes

$$\underline{X}^{\#}(x) = \liminf_{\substack{l \downarrow 0 \\ \gamma \in \Gamma}} \frac{\langle X_{\gamma(l)}, \dot{\gamma}(l) \rangle - \langle X_{\gamma(0)}, \dot{\gamma}(0) \rangle}{\sin l}. \tag{5.22}$$

By inserting $\gamma(0) = x$, $\gamma(l) = y$, (5.20), (5.19), $\langle X_x, x \rangle = \langle X_y, y \rangle = 0$, and $\sin l = \pm\sqrt{1 - \langle x, y \rangle^2}$ into (5.22) we obtain

$$\underline{X}^{\#}(x) = \liminf_{\substack{g \\ y \to x}} \frac{\langle X_y - X_x, y - x \rangle}{1 - \langle x, y \rangle^2}. \tag{5.23}$$

Because $1 - \langle x, y \rangle^2 = (1 - \langle x, y \rangle)(1 + \langle x, y \rangle)$ and $1 + \langle x, y \rangle \to 2$, when $\underset{y \to x}{\overset{g}{}}$, (5.23) becomes

$$\underline{X}^{\#}(x) = \liminf_{\substack{g \\ y \to x}} \frac{\langle X_y - X_x, y - x \rangle}{2 - 2\langle x, y \rangle}. \tag{5.24}$$

But because $\|x\| = \|y\| = 1$ we have that $2 - 2\langle x, y \rangle = \|y - x\|^2$, which inserted into (5.24) gives the required equality of the theorem. $\qquad\square$

For the smooth case we have the following corollary of Theorem 5.22.

COROLLARY 5.23 *If $X = f|_K$, where $f : \mathbb{R}^{n+1} \to \mathbb{R}^{n+1}$ is a smooth operator satisfying $\langle f(x), x \rangle = 0$ when $\|x\| = 1$, then*

$$\underline{X}^{\#}(x) = \inf_{\substack{\|h\|=1, \\ \langle x,h \rangle = 0}} \langle df(x)(h), h \rangle,$$

where $df(x)$ is the Frechét differential of f in x.

Proof. Suppose that $y \to x$ along the arbitrary but fixed geodesic γ starting from x ($\gamma(0) = x$). This means that $y = \gamma(t)$, where $t \downarrow 0$. Hence by using $X = f|_K$ we have that

$$\frac{\langle X_y - X_x, y - x \rangle}{\|y - x\|^2} =$$
$$\left\langle \frac{f(\gamma(t)) - f(\gamma(0))}{t} : \left\|\frac{\gamma(t) - \gamma(0)}{t}\right\|, \frac{\gamma(t) - \gamma(0)}{t} : \left\|\frac{\gamma(t) - \gamma(0)}{t}\right\| \right\rangle \tag{5.25}$$

If $y \to x$ along γ (i.e., $t \downarrow 0$), then by using $(d/dt)|_{t=0}f(\gamma(t)) = df(x)(\dot{\gamma}(0))$, and that γ is of unit speed we obtain from (5.25):

$$\liminf_{\substack{y \to x \\ y \in \text{Im}\gamma}} \frac{\langle X_y - X_x, y - x \rangle}{\|y - x\|^2} = \langle df(x)(\dot{\gamma}(0)), \dot{\gamma}(0) \rangle, \tag{5.26}$$

where we have denoted by $\text{Im}\gamma$ the image of γ. Equation (5.26) holds for every γ, hence by definition of $\underline{X}^{\#}(x)$ we get

$$\underline{X}^{\#}(x) = \inf_{\gamma \in \Gamma} \langle df(x)(\dot{\gamma}(0)), \dot{\gamma}(0) \rangle. \tag{5.27}$$

Because for every $h \in T_x(\mathbb{S}^n)$, $\|h\| = 1$ there is a $\gamma \in \Gamma$ such that $\dot{\gamma}(0) = h$, (5.27) becomes the required equality of the corollary. $\square$

REMARK 5.2 *We use the same notations as above. Then we have that*

1. (a) *Similarly to Theorem 5.22 it can be proved that X is geodesic monotone (strictly geodesic monotone for $K \neq \mathbb{S}^n$) if and only if*

$$\langle X_y - X_x, y - x \rangle \geq 0 \, (\langle X_y - X_x, y - x \rangle > 0),$$

for arbitrary two distinct points x and y of K.

(b) *If X is smooth, by using Theorem 5.18 and Corollary 5.23 we have that X is geodesic monotone if and only if*

$$\langle df(x)(h), h \rangle \geq 0,$$

for all $x \in K$ and all $h \in \mathbb{R}^{n+1} - \{0\}$ with $\langle x, h \rangle = 0$.

(c) *Similarly by using Theorem 5.19 and Corollary 5.23 if X is smooth and*

$$\langle df(x)(h), h \rangle > 0,$$

for all $x \in K \neq \mathbb{S}^n$ and all $h \in \mathbb{R}^{n+1} - \{0\}$ with $\langle x, h \rangle = 0$, then X is strictly geodesic monotone.

2. *If $f = A$, where A is a skew-symmetric operator, then it follows easily that X is trivially geodesic monotone.*

In the following theorem we give a method for computing the scalar derivative of a vector field on the sphere.

THEOREM 5.24 *Let X be a vector field on the sphere, such that $X = f|_{\mathbb{S}^n}$, where $f : \mathbb{R}^{n+1} \to \mathbb{R}^{n+1}$ is a smooth operator satisfying $\langle f(x), x \rangle = 0$, for all $x \in \mathbb{R}^{n+1}$ with $\|x\| = 1$. Let $x \in \mathbb{S}^n$ and let $B(x)$ be an orthogonal matrix having in the first column the components of x relative to the canonical basis of $\mathbb{R}^{n+1}$ and*

$$C(x) = B^T(x) J_s f(x) B(x),$$

where T denotes transposition and

$$J_s f(x) = \frac{Jf(x) + Jf(x)^T}{2}$$

is the symmetrizant of the Jacobi matrix $Jf(x)$ of f in x. Let $C^-(x)$ be the $n \times n$ matrix obtained from $C(x)$ by deleting the first row and first column. Then $\underline{X}^\#(x)$ is the smallest eigenvalue of $C^-(x)$.

Before proving Theorem 5.24 we state a corollary of it.

COROLLARY 5.25 *If $K \subset \mathbb{S}^n$ is open and convex and X is a vector field on K such that $C^-(x)$ is positive definite for all $x \in K$ then X is strictly geodesic monotone.*

The proof of Corollary 5.25 is an easy consequence of Theorems 5.19 and 5.24.

Proof of Theorem 5.24. Corollary 5.23 implies that

$$\underline{X}^{\#}(x) = \inf_{\substack{\|h\|=1 \\ \langle x,h \rangle = 0}} \langle J_s f(x)h, h \rangle.$$

By using the conditional extreme value theorem we obtain

$$\underline{X}^{\#}(x) = \inf\{\lambda : J_s f(x) = \lambda h + \mu x, \text{ for some } \mu \in \mathbb{R}, \text{ and } h \in \mathbb{R}^{n+1}$$
$$\text{with } \|h\| = 1 \text{ and } \langle x, h \rangle = 0\}. \tag{5.28}$$

Consider an orthonormal basis of $\mathbb{R}^{n+1}$ with first vector x. Then in the new basis we have: $h = (0, h_2, \ldots, h_{n+1})$ and the matrix of $d_s f(x) = (df(x) + df(x)^T)/2$ is $C(x)$. If $C(x) = (c_{ij}(x))_{1 \le i,j \le n+1}$, then equation $J_s f(x)h = \lambda h + \mu x$ in the new base is equivalent to

$$\mu = c_{12}h_2 + \cdots + c_{1,n+1}h_{n+1} \quad \text{and} \quad C^-(x)h^- = \lambda h^-, \tag{5.29}$$

where $h^- = (h_2, \ldots, h_{n+1})$. From (5.28) and (5.28) follows immediately the statement of the theorem. $\qquad\square$

Of course similar results can be obtained for positive constant sectional curvature manifolds.

EXAMPLE 5.26 *Let $f : \mathbb{R}^3 \to \mathbb{R}^3$, $f(x) = (x_1^2 - 1, x_1 x_2, x_1 x_3)$, for $x = (x_1, x_2, x_3)$. Because $\langle f(z), z \rangle = 0$ for all $z \in \mathbb{S}^2$ there is a smooth vector field X on $\mathbb{S}^2$ such that $X = f|_{\mathbb{S}^2}$. Hence $X_z = (z_1^2 - 1, z_1 z_2, z_1 z_3)$, for an arbitrary point $z = (z_1, z_2, z_3)$ of $\mathbb{S}^2$. In spherical coordinates ψ, θ we have*

$$\begin{cases} z_1 = \cos\psi \cos\theta, \\ z_2 = \cos\psi \sin\theta, \\ z_3 = \sin\psi. \end{cases}$$

We can choose $B(z)$ to be

$$\begin{pmatrix} \cos\psi\cos\theta & -\sin\theta & \sin\psi\cos\theta \\ \cos\psi\sin\theta & \cos\theta & \sin\psi\sin\theta \\ \sin\psi & 0 & -\cos\psi \end{pmatrix}. \tag{5.30}$$

The Jacobi matrix of f in z is equal to

$$Jf(z) = \begin{pmatrix} 2z_1 & 0 & 0 \\ z_2 & z_1 & 0 \\ z_3 & 0 & z_1 \end{pmatrix}.$$

Hence

$$J_s f(z) = \frac{1}{2} \begin{pmatrix} 4z_1 & z_2 & z_3 \\ z_2 & 2z_1 & 0 \\ z_3 & 0 & 2z_1 \end{pmatrix}.$$

In spherical coordinates this can be written as

$$J_s f(z) = \frac{1}{2} \begin{pmatrix} 4\cos\psi\cos\theta & \cos\psi\sin\theta & \sin\psi \\ \cos\psi\sin\theta & 2\cos\psi\cos\theta & 0 \\ \sin\psi & 0 & 2\cos\psi\cos\theta \end{pmatrix}. \tag{5.31}$$

By using (5.30) and (5.31) we can calculate $C(z) = B^T(z) J_s f(z) B(z)$ in spherical coordinates. Deleting the first row and first column of $C(z)$ we obtain

$$C^-(z) = \begin{pmatrix} \cos\psi\cos\theta & 0 \\ 0 & \cos\psi\cos\theta \end{pmatrix}. \tag{5.32}$$

Equation (5.32) can be written in Cartesian coordinates as

$$C^-(z) = \begin{pmatrix} z_1 & 0 \\ 0 & z_1 \end{pmatrix}. \tag{5.33}$$

By applying Theorem 5.24 to (5.33) we obtain $\underline{X}^{\#}(z) = z_1$. By using Theorem 5.19 the vector field X is strictly geodesic monotone on the hemisphere

$$\begin{cases} z_1^2 + z_2^2 + z_3^2 = 1, \\ z_1 > 0. \end{cases}$$

5.1.7 Geodesic Monotone Vector Fields on $\mathbb{H}^n$

Consider the following model for the n-dimensional hyperbolic space of constant sectional curvature $K = -1$.

$$\mathbb{H}^n = \{\xi = (\xi_1, \ldots, \xi_n, \xi_{n+1}) \in \mathbb{R}^{n+1} : \xi_{n+1} > 0 \quad \text{and} \quad \{\xi, \xi\} = -1\},$$

where for $\xi = (\xi^1, \ldots, \xi^{n+1})$, $\eta = (\eta^1, \ldots, \eta^{n+1}) \in \mathbb{R}^{n+1}$,

$$\{\xi, \eta\} = \xi^1 \eta^1 + \ldots + \xi^n \eta^n - \xi^{n+1} \eta^{n+1}.$$

The metric of $\mathbb{H}^n$ is induced from the Lorentzian metric $\{\cdot, \cdot\}$ of $\mathbb{R}^{n+1}$ and it is denoted by the same symbol. Then an arbitrary geodesic γ of $\mathbb{H}^n$ starting from x ($\gamma(0) = x$) has the equation

$$\gamma(t) = (\cosh t)x + (\sinh t)e, \tag{5.34}$$

where $e = \dot{\gamma}(0) \in T_x(\mathbb{H}^n)$ is the tangent unit vector of γ in the starting point. Also for arbitrary vector field X on $\mathbb{H}^n$ we have

$$\{X_x, x\} = 0, \tag{5.35}$$

for all $x \in \mathbb{H}^n$. If the geodesic γ is passing through $y \in \mathbb{H}^n$ corresponding to parameter l then (5.34) yields

$$y = (\cosh l)x + (\sinh l)e. \tag{5.36}$$

By multiplying (5.36) by x and using the definition of $\mathbb{H}^n$ and (5.35) we obtain

$$\cosh l = -\{y, x\}. \tag{5.37}$$

We also have

$$\cosh^2 t - \sinh^2 t = 1, \tag{5.38}$$

and

$$\liminf_{l \to 0} \frac{\sinh l}{l} = 1. \tag{5.39}$$

By using (5.34) through (5.39) and the definition of $\underline{X}^{\#}(x)$ we can prove, similarly to the previous subsection, the following theorems.

THEOREM 5.27 *Let K be an open and convex subset of $\mathbb{H}^n$, X a vector field on K, and x an arbitrary point of K. Then we have the following formula for the lower geodesic scalar derivative of X in x,*

$$\underline{X}^{\#}(x) = \liminf_{\substack{g \\ y \to x}} \frac{\{X_y - X_x,\, y - x\}}{\{y - x,\, y - x\}},$$

where the limit is taken just for those y which satisfy the relation $\{y-x, y-x\} \neq 0$.

COROLLARY 5.28 *By using the notations of Theorem 5.27 if $X = f|_K$, where $f : \mathbb{R}^{n+1} \to \mathbb{R}^{n+1}$ is a smooth operator satisfying $\{f(x), x\} = 0$ when $\{x, x\} = -1$, then*

$$\underline{X}^{\#}(x) = \inf_{\substack{\{h, h\}=1, \\ \{x, h\}=0}} \{df(x)(h),\, h\},$$

where $df(x)$ is the Frechét differential of f in x.

REMARK 5.3 *We use the same notations as above. Then we have that*

1. (a) *Similarly to Theorem 5.27 it can be proved that X is geodesic monotone (strictly geodesic monotone) if and only if*

$$\{X_y - X_x,\, y - x\} \geq 0 \,(\{X_y - X_x,\, y - x\} > 0),$$

   for arbitrary two distinct points x and y of K.

(b) *If X is smooth, by using Theorem 5.18 and Corollary 5.28 we have that X is geodesic monotone if and only if*

$$\{df(x)(h),\, h\} \geq 0,$$

for all $x \in K$ and all $h \in \mathbb{R}^{n+1} - \{0\}$ with $\{x,\, h\} = 0$ and $\{h, h\} \neq 0$.

(c) *Similarly by using Theorem 5.19 and Corollary 5.28 if X is smooth and*

$$\{df(x)(h), h\} > 0,$$

for all $x \in K$ and all $h \in \mathbb{R}^{n+1} - \{0\}$ with $\{x, h\} = 0$ and $\{h, h\} \neq 0$, then X is strictly geodesic monotone.

2. *If $f = A$, where $A \in o(n,\, 1)$ then it follows easily that X is trivially geodesic monotone.*

We also have an example similar to Example 5.26 of the previous subsection.

EXAMPLE 5.29 *Let*

$$f : \mathbb{R}^3 \to \mathbb{R}^3, \; f(x) = (x_1 x_3, x_2 x_3, x_3^2 - 1),$$

for $x = (x_1, x_2, x_3)$. Inasmuch as

$$\{f(z), z\} = 0,$$

for all $z \in \mathbb{H}^2$ there is a smooth vector field X on H^2 such that

$$X = f|_{\mathbb{H}^2}.$$

Hence

$$X_z = (z_1 z_2, z_2 z_3, z_3^2 - 1),$$

for arbitrary point $z = (z_1, z_2, z_3)$ of $\mathbb{H}^2$. By an easy computation

$$\{df(z)(h), h\} = z_3,$$

for all $z \in \mathbb{H}^2$ and $h \in \mathbb{R}^3$ with $\{h, h\} = 1$, $\{h, z\} = 0$. Hence by using Corollary 5.28 we have that

$$\underline{X}^{\#}(z) = z_3,$$

for all $z \in \mathbb{H}^2$. But $z_3 > 0$, for all $z = (z_1, z_2, z_3) \in \mathbb{H}^2$. Thus, by using Theorem 5.19 X is a strictly geodesic monotone vector field on the whole $\mathbb{H}^2$.

5.2   Killing Monotone Vector Fields

The theory of one-parameter transformation groups is a useful tool for many application-oriented investigations. Many of these results are connected with the qualitative theory of differential equations (see, e.g., [Olver, 1986]), which is very important in theoretical physics. In this section we connect the topic of one-parameter transformation groups with results of nonlinear analysis on Riemannian manifolds, concerning geodesic monotonicity.

We relate the geodesic monotone vector fields to expansive maps. (A map $\phi : M \to M$ is called expansive if its tangential maps are expansive in every point of M.) This is done through the Killing monotone vector fields, a notion introduced by us. The Killing monotone vector fields are vector fields on a Riemannian manifold (M, g), generated by expansive one-parameter transformation groups, where we mean by expansive one-parameter transformation groups smooth one-parameter transformation groups ϕ_t over Riemannian manifolds, with ϕ_t expansive for $t > 0$ (from this follows easily that ϕ_t is nonexpansive for $t < 0$). The Killing strictly monotone vector fields can be introduced similarly. We prove that a vector field X on M is Killing monotone (Killing strictly monotone) if and only if the Lie derivative $\mathcal{L}_X g$, of the metrical tensor g with respect to X is positively semidefinite (positively definite) in every point of M. The positive semidefiniteness (positive definiteness) of $\mathcal{L}_X g$ is proved to be equivalent with the positive semidefiniteness (positive definiteness) of the endomorphism A_X, defined by $A_X U = \nabla_U X$, with respect to g, where ∇ is the Levi–Cività connection of M. This result has two corollaries, regarding the monotonicity of vector fields on Riemannian submanifolds.

Finally, we prove that a vector field is Killing monotone if and only if it is geodesic monotone. From these follows that a necessary and sufficient condition for X to be geodesic monotone is the positive semidefiniteness of $\mathcal{L}_X g$ in every point. We also prove that a necessary condition for X to be strictly geodesic monotone is the positive definiteness of $\mathcal{L}_X g$ in every point of M. We express the lower geodesic scalar derivative of X in terms of the Lie derivative, and we prove that X is geodesic scalar differentiable if and only if it is conformal.

5.2.1   Expansive One-Parameter Transformation Groups

Let (M, g) be a Riemannian manifold.

DEFINITION 5.30 *A smooth map* $\phi : M \to M$ *is called* expansive (nonexpansive), *if for all* $x \in M$ *the tangent map of* ϕ; $T\phi(x) : T_x M \to T_{\phi(x)} M$ *is expansive (nonexpansive); that is,*

$$\|T\phi(x)U\| \geq \|U\| \ (\|T\phi(x)U\| \leq \|U\|) \tag{5.40}$$

for all $U \in T_x M$, *where* $\| \cdot \|$ *is the norm generated by the metric* g.

ϕ *is called* strictly expansive *(strictly nonexpansive) if in (5.40) we have strict inequalities for $U \neq 0$.*

DEFINITION 5.31 *A one-parameter smooth transformation group*

$$\phi_t : M \to M$$

is called expansive (nonexpansive) *if ϕ_t is expansive (nonexpansive) for all $t >$ 0. The* strictly expansive (strictly nonexpansive) *one-parameter transformation groups can be defined similarly.*

DEFINITION 5.32 *Let $\phi_t : M \to M$ be an expansive (nonexpansive) one-parameter transformation group and $X : M \to TM$ the vector field generating ϕ_t that is,*

$$X(z) = \frac{d}{dt}|_{t=0}\phi_t(z),$$

for all $z \in M$. Then the vector field X is called an infinitesimal expansion *(nonexpansion). The* infinitesimal strict expansions *(infinitesimal strict nonexpansions) can be defined similarly. The infinitesimal expansions (infinitesimal strict expansions) are also called* Killing monotone vector fields *(Killing strictly monotone vector fields).*

REMARK 5.4 *The Killing vector fields are trivially Killing monotone vector fields. Hence the Killing monotone vector fields are generalizations of the Killing vector fields.*

The following theorem gives a characterization of the Killing monotone vector fields by using the Lie derivative.

THEOREM 5.33 *$X : M \to TM$ is a Killing monotone vector field (Killing strictly monotone vector field) if and only if (if) the Lie derivative of the metrical tensor $\mathcal{L}_X g$ is positive semidefinite (positive definite) in every point of M.*

Proof. Suppose that X is Killing monotone. Denote $\mathcal{T}M$ the set of all smooth vector fields on M. Let $U, V \in \mathcal{T}M$ be two arbitrary vector fields, $z \in M$ an arbitrary point, and ϕ_t the one-parameter transformation group generated by X. Then we have

$$(\mathcal{L}_X g)(U, V) = X(g(U, V)) - g([X, U], V) - g(U, [X, V]). \quad (5.41)$$

If we put $V = U$ in (5.41), then we obtain

$$(\mathcal{L}_X g)(U, U) = X(g(U, U)) - 2g([X, U], U). \quad (5.42)$$

On the other hand,

$$X(g(U, U))_z = \liminf_{\substack{t \to 0 \\ t > 0}} \frac{g(U(\phi_t(z)), U(\phi_t(z))) - g(U(z), U(z))}{t}. \quad (5.43)$$

Because X is Killing monotone we have that

$$g(U(\phi_t(z)), U(\phi_t(z))) \geq g(T\phi_t^{-1}U(\phi_t(z)), T\phi_t^{-1}U(\phi_t(z))). \tag{5.44}$$

Inserting (5.44) into (5.43) we get

$$X(g(U,U))_z \geq \liminf_{\substack{t \to 0 \\ t > 0}} \left(g\left(\frac{T\phi_t^{-1}U(\phi_t(z)) - U(z)}{t}, T\phi_t^{-1}U(\phi_t(z))\right) \right.$$
$$\left. + g\left(U(z), \frac{T\phi_t^{-1}U(\phi_t(z)) - U(z)}{t}\right) \right). \tag{5.45}$$

But

$$\liminf_{\substack{t \to 0 \\ t > 0}} \frac{T\phi_t^{-1}U(\phi_t(z)) - U(z)}{t} = [X, U]_z. \tag{5.46}$$

Inserting (5.46) into (5.45) we obtain

$$X(g(U,U))_z \geq 2g([X,U], U)_z. \tag{5.47}$$

From (5.42) and (5.47) it follows that $\mathcal{L}_X g$ is positive semidefinite at every $z \in M$.

Conversely, suppose that $\mathcal{L}_X g$ is positive semidefinite (positive definite) at every $z \in M$. We have to prove that $g(T\phi_t U, T\phi_t U) \geq g(U, U)$ (>) for all $t > 0$ and $U \in TM$ ($U \in TM \setminus \{0\}$). But this is trivial because

$$\frac{d}{dt} g(T\phi_t U, T\phi_t U)|_{t=0} = (\mathcal{L}_X g)(U, U). \quad \square$$

Let $X \in TM$ and $A_X : TM \to TM$ the endomorphism defined by

$$A_X U = \nabla_U X,$$

where ∇ is the Levi–Citá connection of M. Then we have the following lemma.

LEMMA 5.34
$$\mathcal{L}_X g(U, U) = 2g(A_X U, U).$$

Proof. Becuase ∇ is the Levi–Citá connection of M we have that

$$X(g(U, U)) = 2g(\nabla_X U, U). \tag{5.48}$$

Inserting (5.48) in (5.42), bearing in mind that ∇ is torsion free and by using the definition of A_X we obtain the required relation. $\quad \square$

By Theorem 5.33 and Lemma 5.34 we have the following theorem.

THEOREM 5.35 $X \in TM$ *is Killing monotone (Killing strictly monotone) if and only if (if) A_X is positive semidefinite (positive definite) relative to the metrical tensor g.*

By using Theorem 5.35 and Corollary 5.20 it is easy to prove the following.

THEOREM 5.36 *Let $K \in M$ be an open and convex set and $X \in TK$. Then we have the following two assertions.*

1. *X is Killing monotone if and only if it is geodesic monotone.*

2. *X is a Killing vector field if and only if it is trivially geodesic monotone.*

Hence the notion of Killing monotone vector fields (Killing vector fields) coincides with those of geodesic monotone (trivially geodesic monotone) ones. We call, for brevity's sake, these vector fields *monotone (trivially monotone)*. However, the relation between strictly geodesic monotone vector fields and Killing strictly monotone vector fields seems to be difficult and we cannot say anything about it yet.

We recall the following definition (see [Kobayashi and Nomizu, 1963, vol. II, p. 53]).

DEFINITION 5.37 *Let M be a Riemannian manifold. A submanifold M of N is called* auto-parallel *if, for each vector $X \in T_x N$ and for each curve τ in N starting from x, the parallel displacement of X along τ (with respect to the affine connection of the ambient space M) yields a vector tangent to N.*

Regarding this notion we recall the following proposition (see [Kobayashi and Nomizu, 1963, vol. II, p. 55], Proposition 8.2).

PROPOSITION 5.4 *Let N be a submanifold of the Riemannian manifold M. Denote ∇ the Levi–Cività connection of M. Then the following two conditions are equivalent.*

1. *N is an auto-parallel submanifold of M.*

2. *If X and Y are vector fields on N, then $\nabla_X Y$ is tangent to N at every point of N.*

Theorem 5.35 and *1* of Theorem 5.36 have the following corollary.

COROLLARY 5.38 *Let (M, g) be a Riemannian manifold, X a geodesic monotone vector field of M, and N an integral submanifold of X. Then the restriction of X to N is a geodesic monotone vector field of N (with respect to the induced metric on N).*

Proof. Denote the restriction of X to N by the same letter. Let $\tilde{\nabla}$ be the Levi–Civitá connection of N (with respect to the induced metric on N) and ∇ be the Levi–Civitá connection of M. Then, by Gauss' formula we have (see [Kobayashi and Nomizu, 1963, vol. II, p. 15 (I)])

$$\nabla_Y X = \tilde{\nabla}_Y X + \alpha(X, Y),$$

where α is the second fundamental form of N and $Y \in TN$. (Y can be extended to M. The extension was denoted by the same letter.) Because $\alpha(X, Y)$ is normal to N we have:

$$g(\nabla_Y X, Y) = g(\tilde{\nabla}_Y X, Y).$$

Because $g(\nabla_Y X, Y) \geq 0$, we get $g(\tilde{\nabla}_Y X, Y) \geq 0$ for all $Y \in TN$. By Theorem 5.35 and *1.* of Theorem 5.36 the restriction of X to N is a geodesic monotone vector field of N. $\square$

EXAMPLE 5.39 *Let* $f : \mathbb{R}^3 \to \mathbb{R}^3$ $f(x, y, z) = (x^3 + x, y + z, z)$. *Then the symmetrizant of the Jacobi matrix of* f *is*

$$df(x, y, z)_s = \begin{bmatrix} x^2 + 1 & 0 & 0 \\ 0 & 1 & \dfrac{1}{2} \\ 0 & \dfrac{1}{2} & 1 \end{bmatrix}.$$

It can be checked that $df(x, y, z)_s$ *is positive definite for all* (x, y, z). *From Theorem* 1.21 *it follows that* f *is a monotone operator. Hence the vector field* X, *where* X_p *is the tangent vector in* $t = 0$ *of the curve* $t \mapsto p + tf(p)$, *is monotone. Consider the cylinder surface* S *given by the equation* $h(x, y, z) = 0$ $z > 0$, *where* $h(x, y, z) = y - z \log z$. *By a straightforward computation it can be verified that*

$$X_x \frac{\partial h}{\partial x} + X_y \frac{\partial h}{\partial y} + X_z \frac{\partial h}{\partial z} = 0,$$

for all $(x, y, z) \in S$. *Hence* S *is an integral surface of* X. *By Corollary* 5.4 *the restriction of* X *to* S *is monotone on* S *with respect to the metric induced from the canonical metric of* $\mathbb{R}^3$.

We also have the following corollary of Theorem 5.35 and 1. of Theorem 5.36 (which generalize Theorem 8.9 of [Ahlfors, 1981, vol. II, p. 59]).

COROLLARY 5.40 *Let* N *be an auto-parallel submanifold of the Riemannian manifold* (M, g). *Let* X *be a monotone vector field of* M. *At each point of* N *decompose* X *into a vector tangent to* N *and a vector normal to* N. *Then the* *tangential component of X is a monotone vector field of N with respect to the induced metric on N.*

Proof. Let $\widetilde{\nabla}$ be the Levi–Civitá connection of N (with respect to the induced metric on N) and ∇ be the Levi–Civitá connection of M. At each point of N we decompose X as follows:

$$X = X' + X'',$$

where X' is tangent to N and X'' is normal to N. Let g also denote the induced Riemannian metric tensor of N. By Theorems 5.35 and 5.36, it suffices to prove that $g(\widetilde{\nabla}_Y X', Y) \geq 0$ for all vector fields Y of N. By the same theorems we have $g(\nabla_Y X, Y) \geq 0$. By Gauss' formula (see [Kobayashi and Nomizu, 1963, vol. II, p. 15 (I)]) we have

$$\nabla_Y X' = \widetilde{\nabla}_Y X' + \alpha(X', Y). \tag{5.49}$$

where α is the second fundamental form of N. Because $\alpha(X', Y)$ is normal to M (5.49) implies

$$0 \leq g(\nabla_Y X, Y) = g(\widetilde{\nabla}_Y X', Y) + g(\nabla_Y X'', Y). \tag{5.50}$$

Because $g(X'', Y) = 0$ and ∇ is the Levi–Civitá connection of N we have

$$g(\nabla_Y X'', Y) = -g(X'', \nabla_Y Y) \tag{5.51}$$

Proposition 5.4 implies

$$g(X'', \nabla_Y Y) = 0. \tag{5.52}$$

Equations (5.51) and (5.52) yield

$$g(\nabla_Y X'', Y) = 0 \tag{5.53}$$

Inserting (5.53) into (5.50) we get $g(\widetilde{\nabla}_Y X', Y) \geq 0$ for all vector fields Y of N. $\square$

EXAMPLE 5.41 *Let $M = \mathbb{R}^n$ and $\langle \cdot, \cdot \rangle$ be the canonical scalar product of $\mathbb{R}^n$. The auto-parallel submanifolds of M are the affine subspaces. Suppose that N is an affine subspace of M. Let $f : \mathbb{R}^n \to \mathbb{R}^n$ be a monotone operator. Then, f induces the monotone vector field X, where X_x is the tangent vector in $t = 0$ of the curve*

$$t \mapsto x + tf(x).$$

In the points of N we can decompose X as follows,

$$X = Y + Z, \tag{5.54}$$

where Y is tangent to N and Z is normal to N. By Corollary 5.40 Y is a monotone vector field on N. We verify this assertion directly. We must prove that $\langle Y_x - Y_y, x - y \rangle \geq 0$ for all $x, y \in N$. Because Z is normal to N and N is an affine subspace of E we have

$$\langle Z_x - Z_y, x - y \rangle = 0. \tag{5.55}$$

Equations (5.54) and (5.55) imply

$$\langle Y_x - Y_y, x - y \rangle = \langle X_x - X_y, x - y \rangle. \tag{5.56}$$

But $\langle X_x - X_y, x - y \rangle \geq 0$, by the monotonicity of X. Hence (5.56) implies

$$\langle Y_x - Y_y, x - y \rangle \geq 0,$$

for any $x, y \in N$.

From Theorem 5.35 and 1 of Corollary 5.20 we get the following result.

THEOREM 5.42 *Let $K \in M$ be an open and convex set and $X \in \mathcal{T}K$. Then X is Killing strictly monotone if $\underline{X}^{\#}(x) > 0$ for all $x \in K$.*

In the previous section we gave examples of strictly geodesic monotone vector fields on the three-dimensional half sphere and on the three-dimensional hyperbolical space. These examples were given by proving the positiveness of the lower scalar derivative . Hence by Theorem 5.42 these vector fields are Killing strictly monotone too.

5.2.2　Geodesic Scalar Derivatives and Conformity

By using Lemma 5.34 and 1 of Corollary 5.20, we easily get the following.

THEOREM 5.43 *Let $X \in \mathcal{T}M$ be a smooth vector field on M and $x \in M$. Then the lower (upper) geodesic scalar derivative of X in x is given by the following formula.*

$$\underline{X}^{\#}(x) = \frac{1}{2} \inf_{g(h,h)=1} (\mathcal{L}_X g)_x(h, h) \; \left(\overline{X}^{\#}(x) = \frac{1}{2} \sup_{g(h,h)=1} (\mathcal{L}_X g)_x(h, h) \right).$$

THEOREM 5.44 *By using the previous notations, if $\mathcal{L}_X g$ is positive definite in every point of K then X is strictly geodesic monotone.*

Proof. Let x be an arbitrary point of K. Because

$$\underline{X}^{\#}(x) = \frac{1}{2} \inf_{g(h,h)=1} (\mathcal{L}_X g)_x(h, h)$$

and the unit ball in T_xM is compact, there is an $h_0 \in T_xM$ with $g(h_0, h_0) = 1$ such that

$$\underline{X}^\#(x) = \frac{1}{2}\mathcal{L}_X g_x(h_0, h_0) > 0.$$

Inasmuch as x has been arbitrarily chosen Theorem 5.19 implies that X is strictly geodesic monotone. □

In the previous section we gave a class of strictly geodesic monotone vector fields, a class which generalizes the notion of position vector fields on an Hadamard manifold.

We recall the following definition.

DEFINITION 5.45 *Let* (M, g) *be a Riemannian manifold and* X *a smooth vector field on* X. *Denote by* $\mathcal{L}_X g$ *the Lie derivative of* g *with respect to* X. *Then* X *is called* conformal *if there is a smooth map* $\lambda : M \to \mathbb{R}$ *such that*

$$\mathcal{L}_X g = \lambda g.$$

THEOREM 5.46 *Let* (M, g) *be a Riemannian manifold and* X *a smooth vector field on* X. *Then* X *is geodesic scalarly differentiable if and only if* X *is conformal.*

Proof. By Theorem 5.43 X is geodesic scalarly differentiable if and only if there is a smooth function $\lambda : M \to \mathbb{R}$ such that

$$\mathcal{L}_X g(Y, Y) = \lambda g(Y, Y) \tag{5.57}$$

for all $Y \in \mathcal{T}M$. From (5.41) it follows that $\mathcal{L}_X g$ is symmetric. But a symmetric bilinear form is determined by its corresponding quadratic form [Gelfand, 1989]. Hence (5.57) implies

$$\mathcal{L}_X g = \lambda g.$$

In addition we have that $\lambda = 2X^\#$. □

5.3   Projection Maps on Hadamard Manifolds

The projection mappings onto closed convex sets of Hilbert spaces are an important tool for many application-oriented investigations. These maps constituted (and still constitute) a great challenge for many mathematicians. The introduction of Kachurovskii–Minty–Browder monotonicity notion (see [Browder, 1964; Kachurovskii, 1960, 1968; Minty, 1962, 1963]) threw a new light on this topic. E. H. Zarantonello's paper [1971] is one of the most important in this subject. In [Zarantonello, 1971] a series of nice monotonicity properties of a projection mappings π_C onto a closed convex subset C of a Hilbert space are proved. We extend one of them, the monotonicity of $I - \pi_C$, to Hadamard manifolds. We introduce a vector field related to a map $f : M \to M$, called the complementary vector field of f, which generalizes the complementary map $I - g$ of a map g from an Euclidean space onto itself. As a further generalization we define the λ-complementary vector field of f. We prove that if f is a projection mappings onto a closed convex set of M and λ increasing, then the λ-complementary vector field of f is monotone. In particular this holds for the complementary vector field too.

5.3.1   Some Basic Consequences of the Comparison Theorems

It is well known that $exp : M \times TM \to M$ is a smooth map. Hence by the definition of the position vector field we have the following proposition.

PROPOSITION 5.5 *Let M be an Hadamard manifold and $\lambda : [0, \infty[\to [0, \infty[$ a function of class $\mathcal{C}^k$ with $\lambda(0) = 0$. Then the map $P^\lambda : M \times M \to TM$ defined by*

$$P^\lambda(x, y) = P_y^{\lambda, x}$$

is of class $\mathcal{C}^k$, where $P^{\lambda, x}$ is the λ-position vector field at x.

In the following definition indices $i = \overline{1, n}$ are considered modulo n.

DEFINITION 5.47 *A geodesic n-sided polygon in a Riemannian manifold M is a set formed by n segments of minimizing unit speed geodesics (called sides of the polygon)*

$$\gamma_i : [0, l_i] \to M; \; i = \overline{1, n},$$

in such a way that $\gamma_i(l_i) = \gamma_{i+1}(0); \; i = \overline{1, n}$. The endpoints of the geodesic segments are called vertices *of the polygon. The angle*

$$\angle(-\dot{\gamma}_i(l_i), \dot{\gamma}_{i+1}(0)); \; i = \overline{1, n}$$

is called the (interior) angle of the corresponding vertex.

Let M be an Hadamard manifold. If a, b, c are three arbitrary points of M then ab will denote the geodesic distance of a from b and $abc_\triangle$ the geodesic triangle of vertices a, b, c (which by the theorem of Hadamard is uniquely defined). In general a geodesic polygon in M, of consecutive vertices $a_1, \dots, a_n$, is denoted $a_1 \dots a_n$.

COROLLARY 5.48 *Let a, b, c, d be the consecutive vertices of a quadrilateral Q in an Hadamard manifold M and $\alpha, \beta, \gamma, \delta$ the angles of the vertices a, b, c, d, respectively. Then*

$$\alpha + \beta + \gamma + \delta < 2\pi.$$

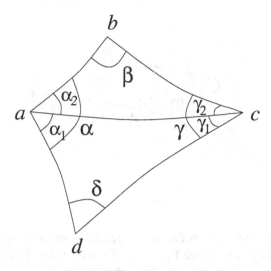

Fig. 5.1

Proof. Let α_1, α_2 be the angles of the vertex a in $adc_\triangle$ and $abc_\triangle$, respectively. Similarly, let γ_1 and γ_2 be the angles of the vertex c in $adc_\triangle$ and $abc_\triangle$, respectively (see Fig. 5.1). It is known that the angle formed by two half straight lines of a trieder is bounded by the sum of the other two angles formed by half straight lines. Hence

$$\alpha_1 + \alpha_2 \geq \alpha, \tag{5.58}$$

and

$$\gamma_1 + \gamma_2 \geq \gamma. \tag{5.59}$$

On the other hand, by 2 of Lemma 5.13 we have

$$\alpha_1 + \gamma_1 + \delta \leq \pi, \tag{5.60}$$

$$\alpha_2 + \gamma_2 + \beta \leq \pi. \tag{5.61}$$

By summing inequalities (5.60), (5.61) and using (5.58), (5.59) we obtain

$$\alpha + \beta + \gamma + \delta \leq 2\pi.$$

$\square$

COROLLARY 5.49 *Let $abc_\triangle$ be a geodesic triangle in an Hadamard manifold M with $A = bc$, $B = ca$, $C = ab$. Let m be the midpoint of the geodesic arc joining b to c, and L_A the length of the geodesic arc joining a to m. Then we have*

$$4L_A^2 \leq 2B^2 + 2C^2 - A^2.$$

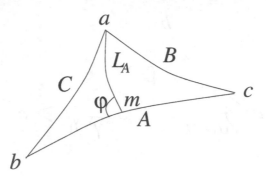

Fig. 5.2

Proof. Let ϕ be the angle between the geodesic arc joining m to a and the geodesic arc joining m to b (see Fig. 5.2). Then by using 1 of Lemma 5.13 to $abm_\triangle$ and $acm_\triangle$, respectively, we obtain the inequalities

$$\frac{A^2}{4} + L_A^2 - AL_A \cos \phi \le B^2, \tag{5.62}$$

and

$$\frac{A^2}{4} + L_A^2 + AL_A \cos \phi \le C^2. \tag{5.63}$$

By summing inequalities (5.62), (5.63) and multiplying by 2 we obtain the required inequality of the corollary. $\qquad\square$

LEMMA 5.50 *Let M be an Hadamard manifold. Let a, b, c, and d be four points of M and let Q be a quadrilateral of consecutive vertices a, b, c, and d. Let α, β, γ, and δ be the angles of the vertices a, b, c, and d, respectively. Suppose that α is nonacute and β is obtuse (nonacute). If $\lambda : [0, \infty[\to [0, \infty[$ is an increasing function, then we have the following assertions.*

1. *$cd > ab$ ($cd \ge ab$).*

2. *$\lambda(da) \cos \delta + \lambda(bc) \cos \gamma \ge 0$.*

   Proof.

1. We consider the following two cases.

   (a) $M = E_2$, where E_2 is the Euclidean plane. Because α is nonacute and β is obtuse (nonacute) the quadrilateral $abcd$ is convex. In this case $\alpha + \beta + \gamma + \delta = 2\pi$. Hence either $\delta \ge \pi - \alpha$ or $\gamma \ge \pi - \beta$. We can suppose without loss of generality that $\delta \ge \pi - \alpha$. Hence the parallel to the straight line ab through d intersects the segment bc in p (see Fig. 5.3).

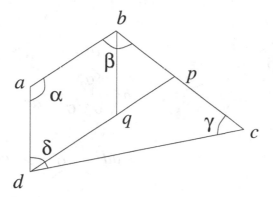

Fig. 5.3

If $p = c$ then $cd > ab$ ($cd \geq ab$) trivially holds. Suppose that $p \neq c$. Then $\widehat{dpc} > \pi/2$ ($\widehat{dpc} \geq \pi/2$). Thus, because the sum of the angles of $dpc_\triangle$ is π, we have that $\widehat{pcd} < \pi/2$ ($\widehat{pcd} \leq \pi/2$). Hence $\widehat{dpc} > \widehat{pcd}$ ($\widehat{dpc} \geq \widehat{pcd}$). This implies that

$$dp < cd \quad (dp \leq cd). \qquad (5.64)$$

Because $\beta \geq \pi/2$, the parallel to the straight line ad through b intersects the segment dp in q (see Fig. 5.3). Then $abqd$ is a parallelogram. Hence,

$$ab = dq \leq dp. \qquad (5.65)$$

Inequalities (5.64) and (5.65) imply that $ab < cd$ ($ab \leq cd$).

(b) In the general case denote by α_1, α_2 the angles of the vertex a in $adc_\triangle$, $abc_\triangle$, respectively (see Fig. 5.4a). It is known that the angle formed by two half straight lines of a trieder is bounded by the sum of the other two angles formed by half straight lines. Hence

$$\alpha_1 + \alpha_2 \geq \alpha. \qquad (5.66)$$

The sides of a geodesic triangle are geodesic arcs of minimal length, therefore the length of a side is bounded by the sum of the lengths of the other two. Hence there is a triangle $ac'd'_\triangle$ in $T_a(M)$, such that $ad' = ad$, $ac' = ac$, and $c'd' = cd$. Similarly in the plane of $ac'd'_\triangle$ there is a point b', such that $ab' = ab$, $b'c' = bc$, and b', d' are in different half-planes of the straight line ac'. Denote the angles of the quadrilateral $ab'c'd'$ corresponding to the vertices a, b', c' d' by α', β', γ', δ', respectively. We denote by α'_1, α'_2 the angles corresponding to the vertex a in $ac'd'_\triangle$, $ab'c'_\triangle$, respectively (see Fig. 5.4b). Then by Lemma 5.13 (i) applied to the $ac'd'_\triangle$ and $ab'c'_\triangle$ we have that

$$\alpha'_1 \geq \alpha_1, \qquad (5.67)$$

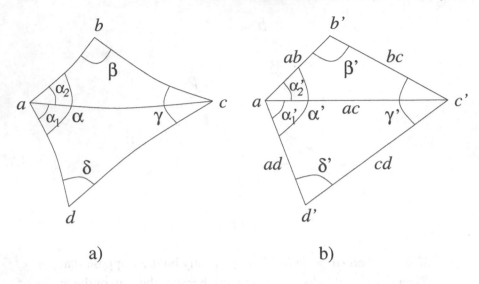

Fig. 5.4

$$\alpha'_2 \geq \alpha_2, \tag{5.68}$$

and

$$\beta' \geq \beta. \tag{5.69}$$

By summing inequalities (5.67), (5.68) and using (5.69) we obtain

$$\alpha' \geq \alpha. \tag{5.70}$$

By inequalities (5.69) and (5.70) the angle α', β' of the quadrilateral $ab'c'd'$ is nonacute, obtuse (nonacute), respectively. Hence by the first case we have that $c'd' \geq ab'$ ($c'd' > ab'$). But $c'd' = cd$ and $ab' = ab$. Hence $cd > ab$ ($cd \geq ab$).

2. By Corollary 5.48 we have $\gamma + \delta \leq \pi$. Hence either $\gamma \leq \pi/2$, or $\delta \leq \pi/2$. Without losing generality we can suppose that $\gamma \leq \delta$. Then we have that

$$\frac{\pi}{2} \geq \frac{\delta + \gamma}{2} \geq \frac{\delta - \gamma}{2} \geq 0. \tag{5.71}$$

Because cos is decreasing on $[0, \pi/2]$ (5.71) implies

$$0 \leq \cos \frac{\delta + \gamma}{2} \leq \cos \frac{\delta - \gamma}{2} \leq 1 \tag{5.72}$$

Inequality (5.72) implies

$$\cos \delta + \cos \gamma = 2 \cos \frac{\delta + \gamma}{2} \cos \frac{\delta - \gamma}{2} \geq 0. \tag{5.73}$$

We consider the following two cases.

(a) $0 < \gamma \leq \delta < \frac{\pi}{2}$. In this case the inequality of 2 is trivial.

(b) $\delta \geq \frac{\pi}{2}$ and $\gamma \leq \frac{\pi}{2}$. Because α, $\delta \geq \pi/2$ by (i) of the lemma we get $0 < da \leq bc$. By using that λ is increasing and $\cos \gamma \geq 0$ we obtain that

$$\lambda(bc) \cos \gamma + \lambda(da) \cos \delta \geq \lambda(da)(\cos \gamma + \cos \delta) \qquad (5.74)$$

By inequalities (5.73) and (5.74) we obtain

$$\lambda(bc) \cos \gamma + \lambda(da) \cos \delta \geq 0.$$

$\square$

The following definition is a particular case of Definition 2.4, Ch. 7 of [do Carmo, 1992].

DEFINITION 5.51 *Let M be an Hadamard manifold and x, $y \in M$. The distance $d(x, y)$ is defined by $d(x, y) =$ the length of the geodesic joining x to y.*

By the Hadamard theorem there is a unique geodesic joining x to y, hence $d(x, y)$ is well defined.

We recall the following proposition (see Proposition 2.5, Ch. 7 of [do Carmo, 1992]):

PROPOSITION 5.6 *With the distance d, M is a metric space; that is:*

1. $d(x, z) \leq d(x, y) + d(y, z)$ *(triangle inequality).*

2. $d(x, y) = d(y, x).$

3. $d(x, y) \geq 0$, *and* $d(x, y) = 0 \Leftrightarrow x = y.$

We recall the following result (see Proposition 2.6, Ch. 7 of [do Carmo, 1992]).

PROPOSITION 5.7 *The topology induced by d on M coincides with the original topology on M.*

DEFINITION 5.52 *Let M be an Hadamard manifold and $o \in M$. If $\widetilde{B}_r(0)$ is the ball of center 0 and radius r in $T_o(M)$, then $\exp_o \widetilde{B}_r(0) = B_r(o)$ is called the geodesic ball with center o and radius r.*

Inasmuch as the exponential map is continuous and $\widetilde{B}_r(0)$ is compact, we have that $B_r(o)$ is compact.

5.3.2   The Complementary Vector Field of a Map

DEFINITION 5.53 *Let M be an Hadamard manifold and $f : M \to M$. Then the vector field $F \in Sec(TM)$ defined by*

$$F_x = P_x^{f(x)}$$

is called the complementary vector field *of f.*

DEFINITION 5.54 *Let M be an Hadamard manifold, $f : M \to M$ and $\lambda : [0, \infty[\to [0, \infty[$. Then the vector field $F \in Sec(TM)$ defined by*

$$F_x^\lambda = P_x^{\lambda, f(x)}$$

is called the λ-complementary vector field of f.

In particular the complementary vector field of f coincides with the id-complementary vector field of f, where "id" is the identity map of M.

By Proposition 5.5 we have the following.

PROPOSITION 5.8 *If $f : M \to M$ and $\lambda : [0, \infty[\to [0, \infty[$ are of class C^k and $\lambda(0) = 0$ then the λ-complementary vector field F^λ of f is of class C^k.*

REMARK 5.5 *Let E be an Euclidean space, $f : E \to E$ and $\lambda : [0, \infty[\to [0, \infty[$. Then*

1. *The complementary vector field F of f assigns to each $x \in E$ the tangent vector in 0 of the curve $t \mapsto x + t(x - f(x))$.*

2. *The λ-complementary vector field F of f assigns to each $x \in E$ the tangent vector in 0 of the curve $t \mapsto x + t\lambda(\|x - f(x)\|)(x - f(x))$.*

5.3.3   Projection maps generating monotone vector fields

DEFINITION 5.55 *Let M be an Hadamard manifold, $x \in M$, and $A \subset M$. Then the nonnegative number*

$$d(x, A) = \inf\{d(x, y)|\ y \in A\}$$

is called the geodesic distance *of x from A.*

Because d is a metric we have the following proposition.

PROPOSITION 5.9 *By using the notations of the previous proposition, we have that the map $d_A; x \mapsto d(x, A)$ is continuous.*

We remark that for a Hadamard manifold M and a closed convex subset A of M d_A is twice differentiable on $M \backslash A$ (see Corollary 1 of [Walter, 1974]).

The following result is a particular case of Theorem 1 of [Walter, 1974]. For the simplicity of the ideas we present a new proof of it.

PROPOSITION 5.10 *Let M be an Hadamard manifold, $x \in M$, and $C \subset M$ a closed convex set. Then the set*

$$\{y \in C| \; d(x,y) = d(x,C)\}$$

contains exactly one element.

Proof. By the definition of $d(x,C)$ there is an $y \in C$ such that

$$d(x,C) \leq d(x,y) < d(x,C) + 1.$$

Hence

$$d(x,C) = d(x,B),$$

where $B = C \cap B_x(d(x,C) + 1)$ and

$$\{y \in C| \; d(x,y) = d(x,C)\} = \{y \in B| \; d(x,y) = d(x,B)\} \qquad (5.75)$$

Because B is a closed subset of the compact set $B_x(d(x,C)+1)$, B is compact. Hence by (5.75) and Proposition 5.9 we have that the set

$$I = \{y \in C| \; d(x,y) = d(x,C)\}$$

is not empty. Now, let us suppose that I contains the elements $p \neq q$. Let u be the midpoint of the geodesic arc joining p to q (see Fig. 5.5). Then by Corollary 5.49 we have that

$$4d(x,u)^2 \leq 2d(x,p)^2 + 2d(x,q)^2 - d(p,q)^2 < 2d(x,p)^2 + 2d(x,q)^2. \quad (5.76)$$

But because $p, q \in M$ we have that

$$d(x,p) = d(x,q) = d(x,C). \qquad (5.77)$$

Inserting (5.77) into (5.76) we obtain

$$d(x,u)^2 < d(x,C)^2. \qquad (5.78)$$

But because C is convex $u \in C$. Hence (5.78) is in contradiction with the definition of $d(x,C)$. Thus I contains exactly one element. $\qquad \square$

The following definition is a particular case of the definition given in Theorem 1 of [Walter, 1974].

DEFINITION 5.56 *Let M be an Hadamard manifold, $x \in M$, and $C \subset M$ a closed convex set. By Proposition 5.10 there is exactly one point $y \in M$ such*

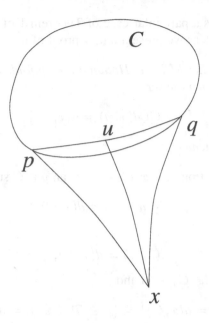

Fig. 5.5

that $d(x, y) = d(x, C)$. *Then* y *is denoted* $\pi_C(x)$ *and is called the* projection *of* x *to* C. *The map* π_C: $x \mapsto \pi_C(x)$ *is called* the projection mappings *to* C.

It is easy to see that $\pi_C^2 = \pi_C$.

The following result generalizes inequality (1.2) of [Zarantonello, 1971] and can be found in [Walter, 1974]. For completeness of the ideas we give a proof of it.

LEMMA 5.57 *Let* (M, g) *be an Hadamard manifold,* $C \subset M$ *a closed convex set,* $x \in M\backslash C$ *and,* $y \in C$. *Then the angle* $\psi(x, y)$ *between the geodesic arc joining* $\pi_C(x)$ *to* x *and the geodesic arc joining* $\pi_C(x)$ *to* y *is nonacute.*

Proof. For p and q in M denote by $\gamma_{pq} : [0, d(p, q)] \to M$ the unit speed geodesic arc joining p to q ($\gamma_{pq}(0) = p$, $\gamma_{pq}(d(p, q)) = q$) and by e_{pq} the unit tangent vector $\dot{\gamma}_{pq}(0)$ of γ_{pq} in 0. Let $y \in C$ and $x \in M\backslash C$. Suppose that $\psi(x, y) < \pi/2$. This is equivalent to

$$g(P^x_{\pi_C(x)}, e_{\pi_C(x)y}) < 0. \tag{5.79}$$

By using (5.79) and the smoothness of P^x we obtain

$$g(P^x_{\gamma_{\pi_C(x)y}(t_0)}, e_{\gamma_{\pi_C(x)y}(t_0)y}) < 0, \tag{5.80}$$

if $t_0 \neq 0$ is sufficiently small. By applying 1 of Lemma 5.13 to the geodesic triangle of vertices x, $\pi_C(x)$ and $\gamma_{\pi_C(x)y}(t_0)$ (see Fig. 5.6) we get

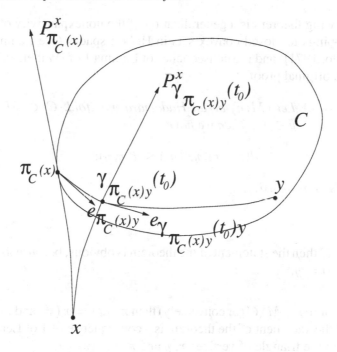

Fig. 5.6

$$d(x, \pi_C(x))^2 \geq d(x, \gamma_{\pi_C(x)y}(t_0))^2 + d(\pi_C(x), \gamma_{\pi_C(x)y}(t_0))^2$$
$$- 2d(\pi_C(x), \gamma_{\pi_C(x)y}(t_0))g(P^x_{\gamma_{\pi_C(x)y}(t_0)}, e_{\gamma_{\pi_C(x)y}(t_0)y}). \tag{5.81}$$

By (5.80) and (5.81) we obtain

$$d(x, C) = d(x, \pi_C(x)) > d(x, \gamma_{\pi_C(x)y}(t_0)). \tag{5.82}$$

But inequality (5.82) is in contradiction with the definition of $d(x, y)$. Hence $\psi(x, y) \geq 0$. □

The following theorem is an easy consequence of Lemma 5.57.

THEOREM 5.58 *Let (M, g) be an Hadamard manifold, $C \subset M$ a closed convex set, $x \in M\backslash C$, and $y \in M$, such that $x \neq y$. Then the angle $\theta(x, y)$ between the geodesic arc joining $\pi_C(x)$ to x and the geodesic arc joining $\pi_C(x)$ to $\pi_C(y)$ is nonacute.*

Theorem 5.58 is a generalization of inequality (1.4) of [Zarantonello, 1971].

We remark that for an Hadamard manifold M the nonexpansivity of a map $\phi : M \to M$ (Definition 5.30) is equivalent to the inequality

$$d(\phi(x), \phi(y)) \leq d(x, y), \quad \forall x \in M.$$

Inequality (5.40) of Definition 5.30 is the local form of the above inequality.

The following theorem is a generalization of the nonexpansivity of the projection mappings to closed convex sets in Hilbert spaces (see Lemma 1.3 of [Zarantonello, 1971]) and is a consequence of Lemma 1 of [Walter, 1974]). We give here an original proof.

THEOREM 5.59 *Let (M, g) be an Hadamard manifold, $C \subset M$ a closed convex set, and $x, y \in M$. Then we have*

$$d(\pi_C(x), \pi_C(y)) \leq d(x, y);$$

that is, π_C is nonexpansive.

Proof.

1. If $x, y \in C$ then the statement of the theorem is obvious, because $\pi_C(x) = x$ and $\pi_C(y) = y$.

2. If $x \in C$ and $y \in M \backslash C$ (or conversely) then $\pi_C(x) = x$ ($\neq$) and $\pi_C(y) \neq y$ (=) and the statement of the theorem is a consequence of 1 of Lemma 5.13 applied to the triangle of vertices x, y and $\pi_C(x)$ ($\pi_C(y)$).

3. If $x, y \in M \backslash C$, then the statement of the theorem is a consequence of 1 of Lemma 5.50 and Theorem 5.58.

$\square$

By Proposition 5.7 Theorem 5.59 has the following corollary (this result can also be found in [Walter, 1974]).

COROLLARY 5.60 *If M is an Hadamard manifold and C a closed convex subset of M, then the projection mappings π_C is continuous on M and differentiable on $M \backslash C$.*

By Proposition 5.8 and Corollary 5.60 we have the following.

PROPOSITION 5.11 *Let M be an Hadamard manifold, C a closed convex subset of M, and $\lambda : [0, \infty[\to [0, \infty[$ a continuous function with $\lambda(0) = 0$. Then the λ-complementary vector field Π_C^λ of the projection map π_C is continuous on M and differentiable on $M \backslash C$.*

PROPOSITION 5.12 *Let (M, g) be an Hadamard manifold, C a closed convex subset of M, and $\lambda : [0, \infty[\to [0, \infty[$ an increasing function. Then the λ-complementary vector field Π_C^λ of the projection mappings π_C is monotone.*

Proof. Let $x \neq y \in M$.

1. If at least one of these points is in C, then by 1 of Lemma 5.13 and Theorem 5.58 we trivially get

$$g(\Pi^\lambda_{C_x}, \dot{\gamma}(0)) \leq g(\Pi^\lambda_{C_y}, \dot{\gamma}(l)), \qquad (5.83)$$

where $\gamma : [0, l] \to M$ is the unit speed geodesic joining x to y, such that $\gamma(0) = x$ and $\gamma(l) = y$ and $l = d(x, y)$.

2. If $x, y \in M \backslash C$, then by Theorem 5.58 and Corollary 5.48 applied to the quadrilateral Q of consecutive vertices $\pi_C(x)$, $\pi_C(y)$, y, and x we have either $\alpha \leq \pi/2$ or $\beta \leq \pi/2$, where α, β are the angles of Q corresponding to the vertices x, y respectively. Then by 2 of Lemma 5.50 we have

$$g(\Pi^\lambda_{C_x}, \dot{\gamma}(0)) \leq g(\Pi^\lambda_{C_y}, \dot{\gamma}(l)). \qquad (5.84)$$

Inequalities (5.83) and (5.84) prove the theorem. $\qquad\qquad\square$

5.4 Nonexpansive Maps

In this section we suppose that M is an Hadamard manifold. If f is (geodesically) nonexpansive, we prove that the complementary vector field of f is monotone. Because a projection mappings onto a closed convex set of an Hadamard manifold is nonexpansive, and compositions of nonexpansive maps are nonexpansive, we can take $f = p_1 \circ \cdots \circ p_n$, where $p_1, \ldots, p_n$ are projection mappings onto closed convex sets of M. For $n = 1$ this is proved in the previous section. In general the composition of projection mappings is not a projection mappings [Zarantonello, 1973], therefore the monotonicity of the complementary vector field of $f = p_1 \circ \cdots \circ p_n$ cannot be reduced to the mentioned result of Section 5.3.

5.4.1 Some Other Consequences of the Comparison Theorems

First we prove the following lemma.

LEMMA 5.61 *Consider $\mathbb{R}^2$ endowed with the canonical scalar product $\langle \cdot, \cdot \rangle$. Let abcd be a quadrilateral in $\mathbb{R}^2$ such that $dc \geq ab$. Denote by $\alpha, \beta, \gamma, \delta$ the angles of the vertices a, b, c, d, respectively. Then*

$$ad \cos \delta + bc \cos \gamma \geq 0. \qquad (5.85)$$

(This holds even if abcd degenerates to a triangle.)

Proof. From $dc \geq ab$ and the Schwartz inequality we have that

$$\langle d - c, a - b \rangle \leq dc^2,$$

which is equivalent to

$$\langle c - d, a - d \rangle + \langle d - c, b - c \rangle \geq 0. \tag{5.86}$$

It is easy to see that (5.86) implies (5.85).   □

The following lemma is a generalization of Lemma 5.61.

LEMMA 5.62 *Let M be an Hadamard manifold and abcd be a quadrilateral in M such that $dc \geq ab$. Denote by $\alpha, \beta, \gamma, \delta$ the angles of the vertices a, b, c, d, respectively. Then*

$$ad \cos \delta + bc \cos \gamma \geq 0.$$

(This holds even if abcd degenerates to a triangle.)

Proof. We identify $T_a M$ with $\mathbb{R}^n$, where $n = \dim M$. If $\delta, \gamma > \pi/2$ then 1 of Lemma 5.50 implies $ab > cd$ which contradicts $cd \geq ab$. Hence we have either $\delta \leq \pi/2$ or $\gamma \leq \pi/2$. We can suppose without loss of generality that

$$\gamma \leq \pi/2. \tag{5.87}$$

The lengths of the sides of a geodesic triangle satisfy the triangle inequalities. Hence there exist the points b', c', d' of $T_a(M)$ such that $ad' = ad$, $ac' = ac$, $d'c' = dc$, $ab' = ab$, $b'c' = bc$, and b' is contained in the plane of $ad'c'_\triangle$, such that b' and d' are contained in different half-planes defined by the straight line in $T_a(M)$ joining a and c'. Let $\alpha' = \angle d'ab'$, $\beta' = \angle ab'c'$, $\gamma' = \angle b'c'd'$, $\delta' = \angle c'd'a$, $\gamma'_1 = \angle ac'd'$, and $\gamma'_2 = \angle ac'b'$ (see Fig. 5.7b). By using Lemma 5.61 to the quadrilateral $ab'c'd'$ we obtain

$$ad' \cos \delta' + b'c' \cos \gamma' \geq 0. \tag{5.88}$$

Denote by γ_1, γ_2 the angles of the vertex c in the triangles $adc_\triangle$, $abc_\triangle$, respectively (see Fig. 5.7a). Then we have, by (i) of Lemma 3.1, p. 259 of [do Carmo, 1992] that

$$\delta' \geq \delta. \tag{5.89}$$

and

$$\gamma'_1 + \gamma'_2 \geq \gamma_1 + \gamma_2 \geq \gamma. \tag{5.90}$$

We consider two cases.

1. $\gamma'_1 + \gamma'_2 \leq \pi$. We have

$$\gamma' = \gamma'_1 + \gamma'_2. \tag{5.91}$$

Relations (5.90) and (5.91) imply

$$\gamma' \geq \gamma. \tag{5.92}$$

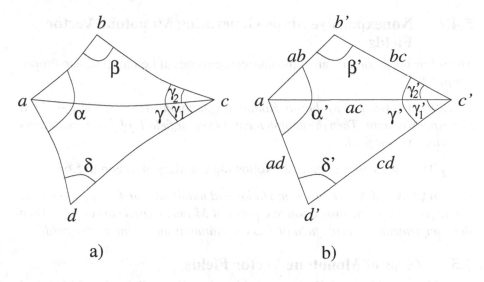

Fig. 5.7

Because $ad' = ad$, $b'c' = bc$ and the cos function is strictly decreasing on $]0, \pi]$ (5.88), (5.89), and (5.92) imply

$$ad \cos \delta + bc \cos \gamma \geq 0.$$

2. $\gamma'_1 + \gamma'_2 > \pi$. If $\delta \leq \pi/2$ then $ad \cos \delta + bc \cos \gamma \geq 0$ holds trivially, because $\gamma \leq \pi/2$. We suppose that $\delta > \pi/2$. By (5.89) we have that $\delta' > \pi/2$. Hence,

$$\gamma'_1 \leq \frac{\pi}{2}. \tag{5.93}$$

We also have

$$\gamma'_2 \leq \pi. \tag{5.94}$$

Relations (5.93) and (5.94) imply

$$2\pi - \gamma' = \gamma'_1 + \gamma'_2 \leq \frac{3\pi}{2}. \tag{5.95}$$

By (5.87) and (5.95) we have $0 < \gamma \leq \gamma' \leq \pi$. Because the cos function is strictly decreasing on $]0, \pi]$ we have

$$\cos \gamma \geq \cos \gamma'. \tag{5.96}$$

Similarly (5.89) implies

$$\cos \delta \geq \cos \delta'. \tag{5.97}$$

By $ad' = ad$, $b'c' = bc$, (5.88), (5.96), and (5.97) we have

$$ad \cos \delta + bc \cos \gamma \geq 0.$$

$\square$

5.4.2 Nonexpansive Maps Generating Monotone Vector Fields

The following theorem is an immediate consequence of Lemma 5.62 and Proposition 5.8.

THEOREM 5.63 *Let M be an Hadamard manifold and $f : M \to M$ a nonexpansive map. Then the complementary vector field F of f is a continuous monotone vector field.*

By Theorem 5.59 we obtain the following corollary of Theorem 5.63.

COROLLARY 5.64 *Let M be an Hadamard manifold and $f = p_1 \circ \cdots \circ p_n$, where $p_1, \cdots, p_n$ are projection mappings of M onto closed convex sets. Then the complementary vector field of f is a continuous monotone vector field.*

5.5 Zeros of Monotone Vector Fields

Let M be a manifold and X a vector field on M. We recall that $a \in M$ is called a *zero* of X if X vanishes at a; that is, $X_a = 0$.

It is known that the set of zeros of a monotone map $A : G \to \mathbb{R}^n$, where G is an open convex subset of $\mathbb{R}^n$, is convex (see, e.g., [Zeidler, 1990], Theorem 32.C.(b)). We generalize this result for Hadamard manifolds.

THEOREM 5.65 *Let (M, g) be an Hadamard manifold, G an open convex set of M, and X a smooth monotone vector field on G. Then the set of zeros of X is convex.*

Proof. If X has at most one fixed point we have nothing to prove. Therefore suppose that X has at least two distinct zeros. Let $a, b \in M$; $a \neq b$ be two arbitrary zeros of X and $\gamma : [0, l] \to M$ be the unit speed geodesic arc beginning at a ($\gamma(0) = a$) and ending at b ($\gamma(1) = b$). We must prove that for all $s \in [0, l]$ $\gamma(s)$ is a zero of X. Suppose that $X_{\gamma(s)} \neq 0$. The monotonicity of X implies that

$$g(X_{\gamma(s)}, \gamma'(s)) = 0. \tag{5.98}$$

Let $\phi_t : M \to M$ be the one-parameter transformation group generated by X. Then by 1. of Theorem 5.36 ϕ_t is nonexpansive for $t < 0$. Because G is open $\gamma(s) \neq \phi_t(\gamma(s))$ and $\phi_t(\gamma(s)) \in G$ for $t < 0$ sufficiently small. Fix such a t_0. By (5.98) the trajectory $\phi_t(\gamma(s))$ is perpendicular to γ. Hence the points a, b, and $c_s = \phi_{t_0}(\gamma(s))$ determine a unique nondegenerate geodesic triangle abc_s. By the triangle inequality we have

$$ac_s + bc_s > ab. \tag{5.99}$$

On the other hand the nonexpansivity of ϕ_{t_0} implies

$$ac_s + bc_s \leq a\gamma(s) + b\gamma(s) = ab. \tag{5.100}$$

But (5.100) is in contradiction with (5.99). Hence we must have $X_{\gamma(s)} = 0$. Thus for all $s \in [0, 1]$ $\gamma(s)$ is a zero of X. $\qquad\square$

5.6 Homeomorphisms and Monotone Vector Fields

Let B be a Banach space and G a subset of B. The map $A : G \to B^*$ is called monotone with respect to duality (or in the sense of Kachurovskii–Minty–Browder [Browder, 1964; Kachurovskii, 1960; Kachurovskii, 1968; Minty, 1962; Minty, 1963]) if $\langle Ay - Ax, y - x \rangle \geq 0$ for any x and y in G, where B^* is the dual of B and $\langle \cdot, \cdot \rangle$ is the natural pairing. If the strict inequality holds whenever $x \neq y$, then A is called strictly monotone. If B is a Hilbert space, then the pairing $\langle \cdot, \cdot \rangle$ can be identified with the scalar product of B. We extended the notion of monotonicity for vector fields of a Riemannian manifold. A classical result of Minty [1962] states that for a Hilbert space H and a continuous monotone map $A : H \to H$, the map $A + I : H \to H$, where I is the identical map of H, is a homeomorphism. This result (and different variations of it) is widely used to prove existence and uniqueness theorems for operator equations, partial differential equations, and variational inequalities (see [Zeidler, 1990]). Surprisingly, in the finite-dimensional case this result boils down just to the continuity and expansivity of $A + I$, being a particular case (it is not trivial to show) of a classical homeomorphism theorem of Browder's Theorem 4.10 [Browder, 1993] (connected to this subject see also [Bae, 1987; Bae and Kang, 1988; Bae and Yie, 1986; Kirk and Schoneberg, 1979; Ray and Walker, 1982, 1985], and [Torrejon, 1983]). We generalize this result for a complete connected Riemannian manifold M. We prove that a continuous expansive map $A : M \to M$ is a homeomorphism. By an expansive map on a Riemannian manifold we mean a map which increases the distance between any two points. The distance function on a Riemannian manifold is given by [do Carmo, 1992, p. 146], Definition 2.4. The expansivity of A can be greatly weakened. It is enough to suppose that A is reverse uniform continuous, which means that for any $\varepsilon > 0$ there is a $\delta = \delta(\varepsilon) > 0$ such that $d(Ax, Ay) < \delta$ implies $d(x, y) < \varepsilon$, where d denotes the distance function on M. Particularly if M is an Hadamard manifold (complete, simply connected Riemannian manifold, of nonpositive sectional curvature) and X is a geodesic monotone vector field on M we prove that $\exp X$ is expansive. Hence if X is continuous $\exp X$ is a homeomorphism of M, extending Minty's classical result. (We note that for a Hilbert space H we have $\exp X = X + I$, where X is identified with a map of H.)

The second author expresses his gratitude to Prof. Tamás Rapcsák, Prof. János Szenthe and Dr. Balázs Csikós for many helpful conversations.

5.6.1   Preliminary Results

First we prove the following lemma.

LEMMA 5.66 *Consider $\mathbb{R}^2$ endowed with the canonical scalar product $\langle \cdot, \cdot \rangle$. Denote by $\| \cdot \|$ the norm induced by $\langle \cdot, \cdot \rangle$. Let abcd be a quadrilateral in $\mathbb{R}^2$ such that $\|c - d\| > \|a - b\|$. Denote by α, β, γ, and δ the angles $\angle dab$, $\angle abc$, $\angle bcd$, and $\angle cda$, respectively. Then,*

$$\|a - d\| \cos \delta + \|b - c\| \cos \gamma > 0. \qquad (5.101)$$

(This holds even if abcd degenerates to a triangle.)

Proof. If $a = b$ the inequality follows from the relation

$$\|a - d\| \cos \delta + \|a - c\| \cos \gamma = \|c - d\|,$$

which can be easily obtained by projecting a to the straight line joining c and d. Suppose that $a \neq b$. From $\|c - d\| > \|a - b\|$ and the Schwartz inequality we have that

$$\langle d - c, a - b \rangle < \|d - c\|^2,$$

which is equivalent to

$$\langle c - d, a - d \rangle + \langle d - c, b - c \rangle > 0. \qquad (5.102)$$

It is easy to see that (5.102) implies (5.101).   □

In the following definition indices $i = 1, \ldots, n$ are considered modulo n. A *geodesic n-sided polygon* in a Riemannian manifold M is a set formed by n segments of minimizing unit speed geodesics (called *sides* of the polygon)

$$\gamma_i : [0, l_i] \to M; \quad i = 1, \ldots, n,$$

in such a way that $\gamma_i(l_i) = \gamma_{i+1}(0); i = l, \ldots, n$. The endpoints of the geodesic segments are called *vertices* of the polygon. The angle

$$\angle(-\dot{\gamma}_i(l_i), \dot{\gamma}_{i+1}(0)); \quad i = 1, \ldots, n$$

is called the (interior) *angle* of the corresponding vertex.

Recall that on Hadamard manifolds every two points can be uniquely joined by a geodesic arc [O'Neil, 1983]. Hence the distance between two points of an Hadamard manifold is the length of the geodesic joining these points.

Let M be an Hadamard manifold. If a, b, c are three arbitrary points of M then ab will denote the distance of a from b and $abc_\triangle$ the geodesic triangle of vertices a, b, c (which is uniquely defined). In general a geodesic polygon in M, of consecutive vertices $a_1, \ldots, a_n$ is denoted $a_1 \ldots a_n$.

LEMMA 5.67 *Let abcd be a quadrilateral in an Hadamard manifold M and* α, β, γ, *and* δ *the angles of the vertices a, b, c, and d, respectively. Then*

$$\alpha + \beta + \gamma + \delta \le 2\pi.$$

Proof. Let α_1 and α_2 be the angles of the vertex a in $adc_\triangle$ and $abc_\triangle$, respectively. Similarly, let γ_1 and γ_2 be the angles of the vertex c in $adc_\triangle$ and $abc_\triangle$, respectively. It is known that an angle formed by two edges of a trieder is bounded by the sum of the other two angles formed by edges. Hence

$$\alpha_1 + \alpha_2 \ge \alpha \tag{5.103}$$

and

$$\gamma_1 + \gamma_2 \ge \gamma. \tag{5.104}$$

On the other hand by [do Carmo, 1992, p. 259], Lemma 3.1 (ii) we have that

$$\alpha_1 + \gamma_1 + \delta \le \pi, \tag{5.105}$$

$$\alpha_2 + \gamma_2 + \beta \le \pi. \tag{5.106}$$

By summing inequalities (5.105), (5.106) and using (5.103), (5.104) we obtain

$$\alpha + \beta + \gamma + \delta \le 2\pi.$$

$\square$

The next lemma follows from Lemma 1 of [Udrişte, 1977].

LEMMA 5.68 *Let M be an Hadamard manifold and abcd be a quadrilateral in M such that α is nonacute and β is obtuse (nonacute), where α, β, γ, δ are the angles of the vertices a, b, c, d, respectively. Then cd > ab (cd $\ge$ ab).*

The following lemma is a generalization of Lemma 5.66.

LEMMA 5.69 *Let M be an Hadamard manifold and abcd be a quadrilateral in M such that cd > ab. Denote by α, β, γ, and δ the angles of the vertices a, b, c, and d, respectively. Then*

$$ad \cos \delta + bc \cos \gamma > 0.$$

(This holds even if abcd degenerates to a triangle.)

Proof. We identify $T_a M$ with $\mathbb{R}^n$, where $n = \dim M$. Denote by $\| \cdot \|$ the norm generated by the canonical scalar product of $\mathbb{R}^n$. If δ, $\gamma \ge \pi/2$ then Lemma 2.3 implies $ab \ge cd$ which contradicts $cd > ab$. Hence we have either $\delta < \pi/2$ or $\gamma < \pi/2$. We can suppose without loss of generality that

$$\gamma < \pi/2. \tag{5.107}$$

The lengths of the sides of a geodesic triangle satisfy the triangle inequalities. Hence there exist the points b', c', d' of $T_a(M)$ such that $\|a - d'\| = ad$, $\|a - c'\| = ac$, $\|d' - c'\| = dc$, $\|a - b'\| = ab$, $\|b' - c'\| = bc$, and b' is contained in the plane of $ad'c'_\triangle$, such that b' and d' are contained in different half-planes defined by the straight line in $T_a(M)$ joining a and c'. Let $\alpha' = \angle d'ab'$, $\beta' = \angle ab'c'$, $\gamma' = \angle b'c'd'$, $\delta' = \angle c'd'a$, $\gamma'_1 = \angle ac'd'$, and $\gamma'_2 = \angle ac'b'$. By using Lemma 5.66 to the quadrilateral $ab'c'd'$ we obtain

$$\|a - d'\| \cos \delta' + \|b' - c'\| \cos \gamma' > 0. \tag{5.108}$$

Denote by γ_1, γ_2 the angles of the vertex c in the triangles $adc_\triangle$, $abc_\triangle$, respectively. Then we have, by [do Carmo, 1992, p. 259], Lemma 3.1 (i) that

$$\delta' \geq \delta \tag{5.109}$$

and

$$\gamma'_1 + \gamma'_2 \geq \gamma_1 + \gamma_2 \geq \gamma. \tag{5.110}$$

We consider two cases:

1. $\gamma'_1 + \gamma'_2 \leq \pi$. We have

$$\gamma' = \gamma'_1 + \gamma'_2. \tag{5.111}$$

   Relations (5.110) and (5.111) implies

$$\gamma' \geq \gamma. \tag{5.112}$$

   Because $\|a - d'\| = ad$, $\|b' - c'\| = bc$, and the cosine function is strictly decreasing on $]0, \pi]$ (5.108), (5.109), and (5.112) imply

$$ad \cos \delta + bc \cos \gamma > 0.$$

2. $\gamma'_1 + \gamma'_2 > \pi$. If $\delta < \pi/2$ then $ad \cos \delta + bc \cos \gamma > 0$ holds trivially, because $\gamma < \pi/2$. We suppose that $\delta \geq \pi/2$. By (5.109) we have that $\delta' \geq \pi/2$. [do Carmo, 1992, p. 259], Lemma 3.1 (ii) implies that

$$\gamma'_1 \leq \pi/2. \tag{5.113}$$

   We also have

$$\gamma'_2 \leq \pi. \tag{5.114}$$

   Hence (5.113) and (5.114) imply

$$2\pi - \gamma' = \gamma'_1 + \gamma'_2 \leq \frac{3\pi}{2}. \tag{5.115}$$

   By (5.107) and (5.115) we have $0 \leq \gamma < \gamma' \leq \pi$. Because the cosine function is strictly decreasing on $[0, \pi]$ we have

$$\cos \gamma > \cos \gamma'. \tag{5.116}$$

Similarly (5.109) implies

$$\cos \delta \geq \cos \delta' \tag{5.117}$$

By $\|a - d'\| = ad$, $\|b' - c'\| = bc$, (5.108), (5.116) and (5.117) we have

$$ad \cos \delta + bc \cos \gamma > 0.$$

$\square$

The next remark follows easily from the definition of a geodesic monotone vector field.

REMARK 5.6 *If (M, g) is an Hadamard manifold, $K \subset M$ a convex open set and $X \in \text{Sec}(TK)$ is a vector field on K, then X is geodesic monotone if and only if for every $x, y \in K$*

$$g(X_x, \exp_x^{-1} y) + g(X_y, \exp_y^{-1} x) \leq 0, \tag{5.118}$$

where $\exp : TM \to M$ is the exponential map of M.

5.6.2 Homeomorphisms of Hadamard Manifolds

The following proposition is a consequence of Lemma 5.69.

PROPOSITION 5.13 *Let (M, g) be an Hadamard manifold and $X \in \text{Sec}(TM)$ a geodesic monotone vector field on M. Then the map $A = \exp X : M \to M$ defined by $Ax = \exp_x X_x$ is expansive.*

Proof. Suppose that A is not expansive. Hence there exist x and y in M such that $x'y' < xy$, where $x' = Ax$ and $y' = Ay$. Consider the quadrilateral $xyy'x'$. Denote by the same letters the angles corresponding to the vertices x and y, respectively. Then by Lemma 5.69 we have

$$xx' \cos x + yy' \cos y > 0 \tag{5.119}$$

It is easy to see that (5.119) is equivalent to

$$g(X_x, \exp_x^{-1} y) + g(X_y, \exp_y^{-1} x) > 0, \tag{5.120}$$

But by (5.118) inequality (5.120) contradicts the monotonicity of X. Hence A is expansive. $\square$

DEFINITION 5.70 *Let M be a Riemannian manifold and d its distance function, which is a metric on M (see [do Carmo, 1992, p. 146], Proposition 2.5). $A : M \to M$ is called* reverse uniform continuous *if for any $\varepsilon > 0$ there is a $\delta = \delta(\varepsilon) > 0$ such that $d(Ax, Ay) < \delta$ implies $d(x, y) < \varepsilon$,*

Let $\alpha \geq 1$ and $L > 0$ be two arbitrary positive constants and $A : M \to M$ such that for any x and y in M we have $d(Ax, Ay) \geq Ld(x, y)^{\alpha}$. Then A is reverse uniform continuous. If $\alpha = L = 1$ we obtain the set of expansive maps.

THEOREM 5.71 *Let M be a complete connected Riemannian manifold and $A : M \to M$ a continuous and reverse uniform continuous map. Then A is a homeomorphism. Particularly this is true for A continuous and expansive.*

Proof. Let $n = \dim M$. It is easy to see that the reverse uniform continuity of A implies that it is injective and $A^{-1} : AM \to M$ is continuous, where $AM = \{Ax : x \in M\}$. Hence $A : M \to AM$ is a homeomorphism. It remains to show that $AM = M$. Suppose that we have already proved that AM is closed. Because $A : M \to AM$ is a homeomorphism, by Brouwer's domain invariance theorem [Massey, 1978, p. 65] AM is open. Because M is connected and AM is an open and closed subset of M we have $AM = M$. Hence if we prove that AM is closed we are done. For this let us consider a sequence $x'_n = Ax_n$ in M convergent to $x' \in M$ and prove that $x' \in AM$ i.e. there is an $x \in M$ such that $x' = Ax$. Because x'_n is convergent it is a Cauchy sequence. It is easy to see that the reverse continuity of A implies that x_n is also a Cauchy sequence. Because M is complete, by the Hopf–Rinow theorem for Riemannian manifolds it is complete as a metric space (see [do Carmo, 1992, p. 146]). Hence x_n is convergent. Denote by x its limit. Because A is continuous taking the limit in the relation $x'_n = Ax_n$ as $n \to \infty$ we obtain $x' = Ax$. $\quad\square$

By Proposition 4.1 we have the following extension to Hadamard manifolds of Minty's classical homeomorphism theorem for monotone maps (Corollary of Theorem 4 of [Minty, 1962]).

COROLLARY 5.72 *Let M be an Hadamard manifold and X be a continuous geodesic monotone vector field. Then $\exp X : M \to M$ is a homeomorphism.*

In Section 5.4 we proved that if $p_1, p_2, \ldots, p_n$ are projection mappings onto closed convex sets of a Hadamard manifold then the vector field

$$X = -\exp^{-1}(p_1 \circ \cdots \circ p_n)$$

defined by

$$X_x = -\exp_x^{-1}[(p_1 \circ \cdots \circ p_n)(x)]$$

is continuous and geodesic monotone. Hence we have the following corollary.

COROLLARY 5.73 *Let M be an Hadamard manifold and $p_1, p_2, \ldots, p_n$ projection mappings onto closed convex sets of M. Then $\exp[-\exp^{-1}(p_1 \circ \cdots \circ p_n)]$ is a homeomorphism of M onto M.*

There are many related nonlinear analysis topics which we did not have space to consider here (see [da Cruz Neto et al., 1999, 2006; Ferreira, 2006; Ferreira and Oliveira, 1998], and [Ferreira et al., 2005]).

References

Ahlfors, L.V. (1981). *Moebius Transformations in Several Dimensions*. Ordway Professorship Lectures in Mathematics. Minneapolis, MN, University of Minnesota, School of Mathematics.

Akhmerov, R.R., Kamenskii, M.I., Potapova, A.S., Rodkina, A.E., and Sadovskii, B.N. (1992). *Measure of Noncompactness and Condensing Operators*. Birkhäuser Verlag, Boston.

Amann, H. (1973). Fixed points of asymptotically linear maps in ordered Banach spaces. *J. Funct. Anal.*, 14:162–171.

Amann, H. (1974a). Lectures on some fixed points theorems. IMPA, Rio de Janeiro.

Amann, H. (1974b). Multiple positive fixed points of asymptotically linear maps. *J. Funct. Anal.*, 17:174–213.

Amann, H. (1976). Fixed points equations and nonlinear eigenvalue problems in ordered Banach spaces. *Siam Rev.*, 18:620–709.

Atiyah, M. F. and Singer, I. M. (1969). Index theory for skew-adjoint Fredholm operators. *Inst. Hautes Études Sci. Publ. Math.*, 37:5–26.

Aussel, D. and Hadjisavvas, N. (2004). On quasimonotone variational inequalities. *J. Optim. Theor. Appl.*, 121(2):445–450.

Bae, J.S. (1987). Mapping theorems for locally expansive operators. *J. Korean Math. Soc.*, 24:65–71.

Bae, J.S. and Kang, B.G. (1988). Homeomorphism theorems for locally expansive operators. *Bull. Korean Math. Soc.*, 25(2):253–260.

Bae, J.S. and Yie, S. (1986). Range of Gateaux differentiable operators and local expansions. *Pacific J. Math.*, 125(2):289–300.

Baiocchi, C. and Capello, A. (1984). *Variational and Quasi-Variational Inequalities. Applications to Free Boundary Problems*. John Wiley and Sons, New York.

Banas, J. and Goebel, K. (1980). *Measure of Noncompactness in Banach Spaces*. Marcel Dekker, New York.

Ben-El-Mechaiekh, H., Chebbi, S., and Florenzano, M. (1994). A Leray–Schauder type theorem for approximable maps. *Proc. Amer. Math. Soc.*, 122(1):105–109.

Ben-El-Mechaiekh, H. and Deguire, P. (1992). Approachability and fixed-points for non-convex set-valued maps. *J. Math. Anal. Appl.*, 170:477–500.

Ben-El-Mechaiekh, H. and Idzik, A. (1998). A Leray–Schauder type theorem for approximable maps: a simple proof. *Proc. Amer. Math. Soc.*, 126(8):2345–2349.

Ben-El-Mechaiekh, H. and Isac, G. (1998). Generalized multivalued variational inequalities. In Cazacu, A., Lehto, O., and Rassias, Th.M., (eds.), *Analysis and Topology*, pages 115–142. World Scientific, Singapore.

Bensoussan, A., Gourset, A., and Lions, J.L. (1973). Côntrole impulsionnel et inéquations quasivariatiônnele stationaires. *C. R. Acad. Sci. Paris, Ser. A-B*, 276:1279–1284.

Bensoussan, A. and Lions, J.L. (1973). Nouvelle formulation des problèmes de contrôle impulsionnel et applications. *C. R. Acad. Sci. Paris, Ser. A-B*, 276:1189–1192.

Berge, C. (1963). *Topological Spaces*. McMillan, New York.

Bianchi, M., Hadjisavvas, N., and Schaible, S. (2004). Minimal coercivity conditions and exceptional families of elements in quasimonotone variational inequalities. *J. Optim. Theor. Appl.*, 122(1):1–17.

Bourbaki, N. (1964). *Éléments de Mathématique, Livre V, Espaces Vectoriel Topologiques, Chaptitres III-V*. Hermann, Paris.

Browder, F. E. (1964). Continuity properties of monotone non-linear operators in Banach spaces. *Bull. Amer. Math. Soc.*, 70:551–553.

Browder, F.E. (1993). Nonlinear operators and nonlinear equations of evolution in Banach spaces. *J. Fac. Sci. Univ. Tokyo Sec. IA*, 40(1):1–16.

Bulavski, V. A., Isac, G., and Kalashnikov, V.V. (1998). Application of topological degree to complementarity problems. In Migdalas, A., Pardalos, P. M., and P. Warbrant (eds.), *Multilevel Optimization: Algorithms and Applications*. Kluwer Academic Publishers. pages 333–358.

Bulavski, V.A., Isac, G., and Kalashnikov, V.V. (2001). Application of topological degree theory to semi-definite complementarity problem. *Optimization*, 49(5-6):405–423.

Burke, J. and Xu, S. (1998). The global linear convergence of a non-interior-point path following algorithm for linear complementarity problems. *Math. Oper. Res.*, 23:719–734.

Burke, J. and Xu, S. (2000). A non-interior predictor-corrector path following algorithm for the monotone linear complementarity problem. *Math. Program.*, 87:113–130.

Cac, N.Ph. and Gatica, J.A. (1979). Fixed point theorems for mappings in ordered Banach spaces. *J. Math. Anal. Appl.*, 71:547–557.

Capuzzo-Dolcetta, J. and Mosco, V. (1980). Implicit complementarity problems and quasivariational inequalities. In Cottle, R.W., Gianessi, F., and Lions, J.L., (eds.), *Variational Inequalities and Complementarity Problems: Theory and Applications*. Wiley, New York.

Carbone, A. and Zabreiko, P.P. (2002). Some remarks on complementarity problems in a Hilbert space. *Anal. Anw.*, 21:1005–1014.

Cellina, A. (1969). A theorem on approximation of compact valued mappings. *Atti Accad. Naz. Lincei Rend. Cl. Sci. Fis. Mat. Nat.*, 47(8):429–443.

Chabat, B. (1991). *Introduction à l'Analyse Complexe*. T.1, Mir Moscou.

Chang, S. S. and Huang, N. J. (1991). Generalized multivalued implicit complementarity problems in Hilbert space. *Mathematica Japonica*, 36:1093–1100.

Chang, S. S. and Huang, N. J. (1993a). Generalized random multivalued quasicomplementarity problems. *Indian J. Math.*, 35:305–320.

Chang, S. S. and Huang, N. J. (1993b). Random generalized set-valued quasicomplementarity problems. *Acta Mathematicae Applicatae Sinica*, 16:396–405.

Chang, S.S. and Huang, N.J. (1993c). Generalized complementarity problems for fuzzy mappings. *Fuzzy Sets Syst.*, 55:227–234.

Chebbi, S. and Florenzano, M. (1995). Maximal elements and equilibra for condensing correspondences. *Cahiers Eco-Math. CERMSEM, Université de Paris I*, 95.18.

Chen, B. and Chen, X. (1999). A global and local superlinear continuation-smoothing method for P_0 and R_0 NCP or monotone NCP. *SIAM J. Optim.*, 9:624–645.

Chen, B., Chen, X., and Kanzow, C. (1997). A generalized Fischer–Burmeister NCP-function: Theoretical investigation and numerical results. Technical report, Zur Angewandten Mathematik, Hamburger Beiträge.

Chen, C. and Mangasarian, O. L. (1996). A class of smoothing functions for nonlinear and mixed complementarity problems. *Comput. Optim. Appl.*, 5:97–138.

Choquet, G. (1969). Outiles topolologiques et metriques de l'analyse mathematiques. (Cours Redige par Claude Mayer) Centre de Documentation Universitaire et S.E.D.E.S., Paris - V.

Cottle, R. W., Pang, J. S., and Stone, R. E. (1992). *The Linear Complementarity Problem*. Academic Press, Boston.

Czerwik, S. (1994). *Stability of Mapping of Hyers–Ulam Type* (Eds. Th.M. Rassias and J. Tabor). Hadronic Press.

Czerwik, S. (2001). *Functional Equations and Inequalities in Several Variable*. World Scientific, New Jersey.

da Cruz Neto, J. X., Ferreira, O. P., and Perez, L. R. Lucambio (1999). A proximal regularization of the steepest descent method in riemannian manifold. *Balkan J. of Geom. Appl.*, (2):1–8.

da Cruz Neto, J.X., Ferreira, O.P., Perez, L.R. Lucambio, and Nemeth, S.Z. (2006). Convex- and monotone-transformable mathematical programming problems and a proximal-like point method. *J. Global. Optim.*, 35(1):53–69.

do Carmo, M.P. (1992). *Riemannian Geometry*. Birkhäuser, Boston.

Facchinei, F. (1998). Structural and stability properties of p_0 nonlinear complementarity problems. *Math. Oper. Res.*, 23:735–749.

Facchinei, F. and Kanzow, C. (1999). Beyond monotonicity in regularization methods for nonlinear complementarity problems. *SIAM J. Control Optim.*, 37:1150–1161.

Ferreira, O.P. (2006). Proximal subgradient and a characterization of lipschitz functions on Riemannian manifolds. *J. Math. Anal. Appl.*, 313(2):587–597.

Ferreira, O.P. and Oliveira, P.R. (1998). Subgradient algorithm on Riemannian manifolds. *J. Optim. Theor. Appl.*, 97(1):93–104.

Ferreira, O.P., Perez, L.R. Lucambio, and Nemeth, S.Z. (2005). Singularities of monotone vector fields and an extragradient-type algorithm. *J. Global. Optim.*, 31(1):133–151.

Ferris, C and Pang, J. S. (1997). Engineering and economic applications of complementarity problems. *SIAM Rev.*, 39:669–713.

Fitzpatrick, P.M. and Petryshin, W.V. (1974). A degree theory, fixed point theorems and mapping theorem for multivalued noncompact mappings. *Trans. Am. Math. Soc.*, 194:1–25.

Gavruta, P. (1994). A generalization of the Hyers–Ulam–Rassias stability of approximately additive mappings. *J. Math. Anal. Appl.*, 184:431–436.

Gelfand, I. M. (1989). *Lectures on Linear Algebra, Translated from the 2nd Russian edition by A. Shenitzer, Reprint of the 1961 translation*. published by Interscience Publishers (English), Dover Books on Advanced Mathematics. New York, Dover.

Georgiev, P. (1997). Submonotone mappings in Banach spaces and applications. *Set-Valued Anal.*, 5(1):1–35.

Giles, J.R. (1967). Classes of semi-inner-product spaces. *Trans. Amer. Math. Soc.*, 129:436–446.

Gorniewicz, L., Granas, A., and Kryszewski, W. (1988/1989). Sur la méthode de l'homotopie dans la théorie des points fixes pour les applications multivoques. Part I: Transveralité topologique. Part II: L'indice dans les ANRS compacts. *C. R. Acad. Sci. Paris Sér. I Math.*, 307/308:489–492/449–452.

Gowda, M.S. and Pang, J.S. (1992). Some existence results for multivalued complementarity problems. *Math. Oper. Res.*, 17:657–670.

Gowda, M.S. and Tawhid, M.A. (1999). Existing and limiting behavior of trajectories associated with p_0-equations. *Comput. Optim. Appl.*, 12:229–251.

Granas, A. (1962). The theory of compact vector fields and some of its applications to topology of functional spaces. *Rozprawy Mathematyczne*, 30:1–93.

Guler, O. (1990). Path following and potential reduction algorithm for nonlinear monotone complementarity problems. Technical report, Department of Management Sciences, The University of Iowa, Iowa city.

Guler, O. (1993). Existence of interior points and interior-point paths in nonlinear monotone complementarity problems. *Math. Oper. Res.*, 18:128–147.

Hamburg, P., Mocanu, P., and Negoescu, N. (1982). *Analiză Matematică (Funcţii Complexe)*. Ed. Didactică şi pedagogică, Bucureşti.

Harker, P.T. and Pang, J.S. (1990). Finite-dimensional variational inequality and nonlinear complementarity problems: A survey of theory, algorithms and applications. *Math. Program.*, 48:161–220.

Hotta, K. and Yoshise, A. (1999). Global convergence of a class of non-interior-point algorithms using Chen–Harker–Kanzow functions for nonlinear complementarity problems. *Math. Program.*, 86:105–133.

Huang, N. J. (1998). A new class of random completely generalized strongly nonlinear quasi-complementarity problems for random fuzzy mappings. *The Korean J. Comput. & Appl. Math.*, 5(2):315–329.

Hyers, D. H., Isac, G., and Rassias, Th.M. (1998a). *Stability of Functional Equations in Several Variables*. Birkhäuser, Boston.

Hyers, D. H., Isac, G., and Rassis, T. M. (1997). *Topics in Nonlinear Analysis and Applications*. World Scientific, Singapore.

Hyers, D.H. (1941). On the solvabiliy of the linear functional equation. *Proc. Nat. Acad. Sci. U.S.A*, 27:222–224.

Hyers, D.H., Isac, G., and Rassias, Th.M. (1998b). On the asymptoticity aspect of Hyers–Ulam stability of mapping. *Proc. Amer. Math. Soc.*, 126(2):425–430.

Isac, G. (1982). Opérateurs asymptotiquement linéaires sur des espaces localement convexes. *Colloquium Math.*, XLVI(1):67–72.

Isac, G. (1986). Complementarity problem and coincidence equations on convex cones. *Bollettino della Unione Matematica Italiana B*, 6(5):925–943.

Isac, G. (1990). A special variational inequality and implicit complementarity problems. *J. Fac. Sci. Univ. Tokyo Sect. IA, Math.*, 37:109–127.

Isac, G. (1992). *Complementarity Problems*. LNM 1528. Springer-Verlag, New York.

Isac, G. (1999a). A generalization of karamardian's condition in complementarity theory. *Nonlinear Anal. Forum*, 4:49–63.

Isac, G. (1999b). On the solvability of multivalued complementarity problem: A topological method. In Felix, R., (ed.), *Fourth European Workshop on Fuzzy Decision Analysis and Recognition Technology, EFDAN"99*. Dorthmund, Germany, pages 51–66.

Isac, G. (1999c). The scalar asymptotic derivative and the fixed point theory on cones. *Nonlinear Analysis and Related Topics (National Academy of Sciences of Belarus, Proceedings of the Institute of Mathematics)*, 2:92–97.

Isac, G. (2000a). $(0, k)$-epi mappings. Applications to complementarity theory. *Math. Comput. Model.*, 32:1433–1444.

Isac, G. (2000b). Exceptional family of elements, feasibility, solvability and continuous path of ε-solutions for nonlinear complementarity problems. In Pardalos, P.M., (ed.), *Approximation and Complexity in Numerical Optimization: Continuous and Discrete Problems*, pages 323–337. Kluwer Academic, Boston.

Isac, G. (2000c). On the solvability of complementarity problems by leray–schauder type alternatives. *Libertas Mathematica*, 20:15–22.

Isac, G. (2000d). *Topological Methods in Complementarity Theory*. Kluwer Academic, Boston.

Isac, G. (2001). Leray–Schauder type alternatives and the solvability of complementarity problems. *Topolog. Meth. Nonlinear Anal.*, 18:191–204.

Isac, G. (2006). *Leray–Schauder type alternatives, Complementarity Problems and Variational Inequalities*. Springer-Verlag, New York.

Isac, G., Bulavski, V. A., and Kalashnikov, V.V. (1997). Exceptional families, topological degree and complementarity problems. *J. Global. Optim.*, 10(2):207–225.

Isac, G., Bulavski, V. A., and Kalashnikov, V.V. (2002). *Complementarity, Equilibrium, Efficiency and Economics*. Kluwer Academic.

Isac, G. and Carbone, A. (1999). Exceptional families of elements for continuous functions: Some applications to complementarity theory. *J. Global. Optim.*, 15(2):181–196.

Isac, G. and Cojocaru, M. G. (2002). Functions without exceptional family of elements and the solvability of variational inequalities on unbounded sets. *Topolog. Meth. Nonlinear Anal.*, 20:375–391.

Isac, G. and Goeleven, D. (1993a). Existence theorems for the implicit complementarity problems. *Int. J. Math., Math. Sci.*, 16:67–74.

Isac, G. and Goeleven, D. (1993b). The implicit general order complementarity problem: Models and iterative methods. *Ann. Oper. Res.*, 44:63–92.

Isac, G. and Gowda, M. S. (1993). Operators of class $(s)_+^1$, Altman's condition and the complementarity problems. *J. Fac. Sci. The Univ. Tokyo Sect. IA*, 40(1):1–16.

Isac, G. and Kalashnikov, V.V. (2001). Exceptional families of elements, Leray–Schauder alternative, pseudomonotone operators and complementarity. *J. Optim. Theor. Appl.*, 109(1):69–83.

Isac, G., Kostreva, M.M., and Polyashuk, M. (2001). *Relational complementarity problem, From local to Global Optimization*, pages 327–339. Kluwer Academic, Boston, A. Migdalas et al. edition.

Isac, G. and Nemeth, S.Z. (2003). Scalar derivatives and scalar asymptotic derivatives. properties and some applications. *J. Math. Anal. Appl.*, 278(1):149–170.

Isac, G. and Nemeth, S.Z. (2004). Scalar derivatives and scalar asymptotic derivatives. an Altman type fixed point theorem on convex cones and some applications. *J. Math. Anal. Appl.*, 290(2):452–468.

Isac, G. and Nemeth, S.Z. (2005a). Duality in multivalued complementarity theory by using inversions and scalar derivatives. *J. Global. Optim.*, 33(2):197–213.

Isac, G. and Nemeth, S.Z. (2005b). Linear complementarity problems and matrix factorization. Research report, Department of Mathematics and Computer Science, Royal Military College of Canada, Kingston, Ont., Canada, February.

Isac, G. and Nemeth, S.Z. (2006a). Duality in nonlinear complementarity theory by using inversions and scalar derivatives. *Math. Inequal. Appl.*, 9(4).

Isac, G. and Nemeth, S.Z. (2006b). Duality of implicit complementarity problems by using inversions and scalar derivatives. *J. Optim. Theor. Appl.*, 128(3):621–633.

Isac, G. and Nemeth, S.Z. (2006c). Fixed points and positive eigenvalues for nonlinear operators. *J. Math. Anal. Appl.*, 314(2):500–512.

Isac, G. and Nemeth, S.Z. (2007a). Browder–Hartman–Stampacchia Theorem and the existence of a bounded interior band of ε-solutions for nonlinear complementarity problems. *Rocky Mountain J. Math.* (accepted).

Isac, G. and Nemeth, S.Z. (2007b). REFE-acceptable mappings: A necessary and sufficient condition for the nonexistence of a regular exceptional family of elements. *J. Optim. Theor. Appl.* (accepted).

Isac, G. and Obuchowska, W.T. (1998). Functions without exceptional families of elements and complementarity problems. *J. Optim. Theor. Appl.*, 99(1):147–163.

Isac, G. and Rassias, Th.M. (1993a). On the Hyers–Ulam stability of ψ-additive mappings. *J. Approx. Theor.*, 72(2):131–137.

Isac, G. and Rassias, Th.M. (1993b). Stability of ψ-additive mappings: Applications to nonlinear analysis. *J. Approx. Theor.*, 19(2):131–127.

Isac, G. and Rassias, Th.M. (1994). Functional inequalities for approximately additive mappings. In Rassias, Th.M. and Tabor, J. (eds.), *Stability of mappings of Hyers–Ulam Type*, pages 117–125. Hadronic Press.

Isac, G. and Zhao, Y.B. (2000). Exceptional families of elements and the solvability of variational inequalities for unbounded sets in infinite dimensional Hilbert spaces. *J. Math. Anal. Appl.*, 246(2):544–556.

Kachurovskii, R.I. (1960). On monotone operators and convex functionals. *Uspekhi. Mat. Nauk*, 15(4):213–215.

Kachurovskii, R.I. (1968). Nonlinear monotone operators in Banach spaces (russian). *Uspekhi. Mat. Nauk (N.S.)*, 23(2):121–168.

Kachurovskii, R. I. (1960). On monotone operators and convex functionals. *Uspekhi Mat. Nauk SSSR*, 15(4):213–215 (Russian).

Kalashnikov, V.V. and Isac, G. (2002). Solvability of implicit complementarity problems. *Annals Oper. Res.*, 116:199–221.

Kantorovici, I. V. and Akilov, G. P. (1977). *Functional Analysis (in Russian)*. Nauka, Moscow.

Kanzow, C. (1996). Some nonlinear continuation methods methods for nonlinear complementarity problems. *SIAM J. Matrix Anal. Appl.*, 17:851–868.

Karamardian, S. and Schaible, S. (1990). Seven kinds of monotone maps. *J. Optim. Theor. Appl.*, 66(1):37–46.

Kinderlehrer, D. and Stampacchia, G. (1980). *An Introduction to Variational Inequalities and their Applications*. Academic Press, New York.

Kirk, W. A. and Schoneberg, R. (1979). Mapping theorems for local expansions in metric and Banach spaces. *J. Math. Anal. Appl.*, 72:114–121.

Klingenberg, W. (1978). *Lectures on Closed Geodesics*. Springer-Verlag, New-York.

Kobayashi, S. and Nomizu, K. (1963). *Foundations of Differential Geometry, Vol. I, II*. Interscience Publishers, New-York.

Kojima, M., Megiddo, N., and Noma, T. (1991a). Homotopy continuation methods for nonlinear complementarity problems. *Math. Oper. Res.*, 16:754–774.

Kojima, M., Megiddo, N., Noma, T., and Yoshise, A. (1991b). *A unified Approach to Interior Point Algorithms for Linear Complementarity Problems*. LNCS 538. Springer-Verlag, New York.

Krasnoselskii, M.A. (1964a). *Positive Solutions of Operator Equations*. P. Noordhoff, Groningen.

Krasnoselskii, M.A. (1964b). *Topological Methods in the Theory of Nonlinear Integral Equations*. Pergamon Press, Oxford.

Krasnoselskii, M.A. and Zabreiko, P. P (1984). *Geometrical Methods of Nonlinear Analysis*. Springer-Verlag, New York.

Lumer, G. (1961). Semi-inner-product spaces. *Trans. Amer. Math. Soc.*, 100:29–43.

Luna, G. (1975). A remark on the nonlinear complementarity problem. *Proc. Amer. Math. Soc.*, 48(1):132–134.

Marinescu, G. (1963). *Espace Vectoriels Pseudotopologiques et Théorie des Distributions*. Hochschulbücher für Mathematik, Band 59 VEB Deutscher Verlag der Wissenschaften, Berlin.

Massey, W.S. (1978). *Homology and Cohomology Theory, An Approach Based on Alexander–Spanier Cochains*, volume 46 of *Monographs and Textbooks in Pure and Applied Mathematics*. Marcel Dekker, New York.

Megiddo, N. (1989). Pathways to the optimal set in nonlinear programming. In Megiddo, N., (ed.), *Progress in Mathematical Programming: Interior-Point and Related Methods*, pages 131–158. Springer-Verlag, New York.

Mininni, M. (1977). Coincidence degree and solvability of some nonlinear functional equations in normed spaces: A spectral approach. *Nonlinear Anal. Theor. Meth. Appl.*, 1(2):105–122.

Minty, G. (1962). Monotone operators in Hilbert spaces. *Duke Math. J.*, 29:341–346.

Minty, G. (1963). On a "monotonicity" method for the solution of non-linear equations in Banach spaces. *Proc. Nat. Acad. Sci. USA*, 50:1038 – 1041.

Monteiro, R. D. C. and Adler, I. (1989). Interior path following primal dual algorithms, Part I: Linear Programming. *Math. Program.*, 44:27–42.

Mosco, U. (1976). *Implicit Variational Problems and Quasi-Variational Inequalities*, volume 543. LNM 543. Springer-Verlag, New York.

Mosco, U. (1980). On some nonlinear quasi-variational inequalities and implicit complementarity problems in stochastic control theory. In Cottle, R. W., Gianessi, F,. and Lions, J.L. (eds.), *Variational Inequalities and Complementarity Problems Theory and Applications*. Wiley, New York.

Nash, C. and Sen, S. (1983). *Topology and Geometry for Physicists*. Academic Press, San Diego.

Nemeth, S.Z. (1992). A scalar derivative for vector functions. *Rivista di Matematica Pura ed Applicata*, (10):7–24.

Nemeth, S.Z. (1993). Scalar derivatives and spectral theory. *Mathematica*, Thome 35(88), N1: 49–58.

Nemeth, S.Z. (1997). Scalar derivatives and conformity. *Annales Univ. Sci. Budapest*, 40: 105–111.

Nemeth, S.Z. (1999a). Geodesic monotone vector fields. *Lobachevskii J. Math.*, 5:13–28.

Nemeth, S.Z. (1999b). Monotone vector fields. *Publicationes Mathematicae Debrecen*, 54(3-4):437–449.

Nemeth, S.Z. (1999c). Monotonicity of the complementarity vector field of a nonexpansive map. *Acta Mathematica Hungarica*, 84(3):189–197.

Nemeth, S.Z. (2001). Homeomorphisms and monotone vector fields. *Publicationes Mathematicae Debrecen*, 58(4):707–716.

Nemeth, S.Z. (2006). Scalar derivatives in Hilbert spaces. *Positivity*, 10(2):299–314.

Nussbaum, R.D. (1971). The fixed point index for local condensing maps. *Ann. Mat. Pura Appl.*, 84:217–258.

Obuchowska, W.T. (2001). Exceptional families and existence results for nonlinear complementarity problem. *J. Global. Optim.*, 19(2):183–198.

Olver, P.J. (1986). *Applications of Lie Groups to Differential Equations*. Springer-Verlag, New York.

O'Neil, B. (1983). *Semi-Riemannian Geometry. With Applications to Relativity*. Pure and Applied Mathematics, 106. Academic Press XIII, New York.

Pang, J. S. (1995). *Complementarity Problems, Horst, R. and Pardalos, P.M (Eds.), Handbook of Global Optimization*, pages 271–338. Kluwer Academic, Boston.

Parida, J. and Sen, A. (1987). A class of nonlinear complementarity problems for multifunctions. *J. Optim. Theor. Appl.*, 53:105–113.

Radu, V. (2003). The fixed point alternative and the stability of functional equations. *Fixed Point Theory*, 4(1):91–96.

Rapcsak, T. (1997). *Smooth Nonlinear Optimization in* $\mathbf{R}^n$. Kluwer Academic, Boston.

Rassias, Th.M. (1978). On the solvability of the linear mapping in Banach spaces. *Proc. Amer. Math. Soc.*, 72:279–300.

Ray, W.O. and Walker, A.M. (1982). Mapping theorems for Gateaux differentiable and accretive operators. *Nonlinear Anal. Theor. Meth. Appl.*, 6(5):423–433.

Ray, W.O. and Walker, A.M. (1985). Perturbations of normally solvable nonlinear operators. *Int. J. Math. Math. Sci.*, 8:241–246.

Riesz, F. and Nagy, B.Sz. (1990). *Functional Analysis, Translated from the second French edition by Leo F. Boron. Reprint of the 1955 original*. Dover Books on Advanced Mathematics. New York.

Sadovskii, B.N. (1968). On measure of noncompactness and concentrative operators. *Probl. Mat. Analiza Sloz. Sistem, Voronez*, 2:89–119.

Sadovskii, B.N. (1972). Limit compact and condensing operators. *Russian Math. Surveys*, 27(1): 85–155.

Saigal, R. (1976). Extension of the generalized complementarity problem. *Math. Oper. Res*, 1(3):260–266.

Spingarn, J. E. (1981). Submonotone subdifferentials of Lipschitz functions. *Trans. Amer. Math. Soc.*, 264:77–89.

Spivak, M. (1965). *Calculus on Manifolds. A Modern Approach to Classical Theorems of Advanced Calculus*. W.A. Benjamin, New York.

Talman, L. A. (1973). Fixed point theorems for mappings in ordered Banach spaces. PhD Thesis, B.A. College of Wooster, University of Kansas.

Torrejon, R. (1983). A note on locally expansive and locally accretive operators. *Canad. Math. Bull.*, 26(2):228–232.

Tseng, P. (1997). An infeasible path-following method for monotone complementarity problems. *SIAM J. Optim.*, 7(2):386–402.

Udriște, C. (1976). Convex functions on Riemannian manifolds. *St. Cerc. Mat.*, 28(6):735–745.

Udriște, C. (1977). Continuity of convex functions on Riemannian manifolds. *Bull. Math. Roum.*, 21:215–218.

Walter, R. (1974). On the metric projections onto convex sets in Riemannian spaces. *Arch. Math.*, 25:91–98.

Wojtaszczyk, P. (1991). *Banach Spaces for Analysts*. Cambridge University Press, New York.

Zabreiko, P.P., Koshelev, A.I., Krasnoselskii, M.A., Mikhlin, S.G., Rakovshchik, L.S., and Stetsenko, V.Ya (1975). *Integral Equations. A Reference Text*. Nordhoff International.

Zarantonello, E. H. (1973). Proyecciones sobre conjuntos convexos en el espacio de Hilbert y teoria espectral. *Revista de la Unión Mathemática Argentina*, 26:187–201.

Zarantonello, E.H. (1971). Projections on convex sets in Hilbert space and spectral theory, I: Projections on convex sets, II: Spectral theory. *Contrib. Nonlin. Functional Analysis, Proc. Sympos. Univ. Wisconsin, Madison*, pages 237–424.

Zeidler, E. (1986). *Nonlinear Functional Analysis and its Applications, Part 1: Fixed Point Theorems*. Springer-Verlag, New York.

Zeidler, E. (1990). *Nonlinear Functional Analysis and its Applications II/B: Nonlinear Monotone Operators*. Springer-Verlag, New York.

Zhao, Y.B. (1998). *Existence Theory and Algorithms for Finite-Dimensional Variational Inequality and Complementarity Problems*. PhD thesis, Academia SINICA, Beijing.

Zhao, Y.B. (1999). *d*-orientation sequences for continuous and nonlinear complementarity problems. *Appl. Math. and Comput.*, 106(2–3):221–235.

Zhao, Y.B. and Han, J. Y. (1999). Exceptional family of elements for variational inequality problem and its applications. *J. Global. Optim.*, 14(3):313–330.

Zhao, Y.B., Han, J. Y., and Qi, H. D. (1999). Exceptional families and existence theorems for variational inequality problems. *J. Optim. Theor. Appl.*, 101(2):475–495.

Zhao, Y.B. and Isac, G. (2000a). Properties of multivalued mapping associated with some non-monotone complementarity problems. *SIAM J. Control Optim.*, 39(3):571–593.

Zhao, Y.B. and Isac, G. (2000b). Quasi-P_*-maps and $P(\tau, \alpha, \beta)$-maps, exceptional family of elements and complementarity problems. *J. Optim. Theor. Appl.*, 105(1):213–231.

Zhao, Y.B. and Li, D. (2000). Strict feasibility conditions in nonlinear complementarity problems. *J. Optim. Theor. Appl.*, 107(2):641–664.

Zhao, Y.B. and Li, D. (2001a). Characterizations of a homotopy solution mapping for quasi-and semi-monotone complementarity problems. Preprint, Academia SINCA, Beijing, (China).

Zhao, Y.B. and Li, D. (2001b). On a new homotopy continuation trajectory for nonlinear complementarity problems. *Math. Oper. Res*, 26(1):119–146.

Zhao, Y.B. and Sun, D. (2001). Alternative theorems for nonlinear projection equations and applications. *Nonlinear Analysis, Theory Methods & Applications*, 46A(6):853–868.

Zhao, Y.B. and Yuan, J. Y. (2000). An alternative theorem for generalized variational inequalities and solvability of non-linear quasi-P_*^M-complementarity problems. *Appl. Math. Comput.*, 109(2–3):167–182.

Index